SIX SIGMA DEPLOYMENT

SIX SIGMA DEPLOYMENT

Cary W. Adams
Praveen Gupta
Charles E. Wilson, Jr.

An imprint of Elsevier Science
www.bh.com

Amsterdam   Boston   London   New York   Oxford   Paris
San Diego   San Francisco   Singapore   Sydney   Tokyo

Butterworth-Heinemann is an imprint of Elsevier Science.

∞ Recognizing the importance of preserving what has been written, Elsevier Science prints its books on acid-free paper whenever possible.

Library of Congress Cataloging-in-Publication Data

Adams, Cary W., 1946-
  Six Sigma deployment / Cary W. Adams, Praveen Gupta, Charles E. Wilson, Jr.
    p. cm.
  Includes index.
  ISBN 0-7506-7523-3
    1. Quality control—Statistical methods. 2. Production management—Statistical methods. I. Gupta, Praveen II. Wilson, Charles E. III. Title.

TS156 .G865 2002
658.5$'$62–dc21   2002035624

British Library Cataloguing-in-Publication Data
A catalogue record for this book is available from the British Library.

The publisher offers special discounts on bulk orders of this book.
For information, please contact:

Manager of Special Sales
Elsevier Science
200 Wheeler Road
Burlington, MA 01803
Tel: 781-313-4700
Fax: 781-313-4880

For information on all Butterworth-Heinemann publications available, contact our World Wide Web home page at: http://www.bh.com

10 9 8 7 6 5 4 3 2 1

Printed in the United States of America

CONTENTS

FOREWORD

Six Sigma has become a proven methodology and strategic initiative that corporations are deploying to realize dramatic benefits. Motorola, GE, Honeywell, Polaroid, ABB, and many more companies have reported economic gains by implementing Six Sigma. Yet, many executives are still wondering about Six Sigma. To them it is like a myth—some fancy program that is for large companies only. They think it is another "program of the year," and they are tired of such programs.

Understanding Six Sigma is a first step in order for an executive or engineer to implement it successfully. Several books have been published on the subject, each focusing on certain aspects of Six Sigma. *Six Sigma Deployment* really takes us back to the origin of Six Sigma and explains how it evolved over the years; it is quite refreshing.

Many people have been misled into thinking of Six Sigma as just a statistical approach to problem analysis. It is much more than that; it is a culture—a way of thinking, a set of new behaviors—that is unique unto itself. The Six Sigma concept is based upon a time-proven approach that has evolved over the past 75 years. Its foundation is Shewhart's groundbreaking work in the 1920s. It went on to incorporate the recent work done by the best organizational behaviorists, thereby becoming a holistic approach to quality, goal setting, and problem analysis. It provides any organization with a means to significantly improve both the soft and hard measurements that are so critical to business success in today's multicultural, highly competitive environment.

The authors of this book have attempted to differentiate it from other books in the way that it describes key aspects of the current Six Sigma methodology, such as strategic plan, project selection, Black Belt training, and project management. This is a unique presentation of the Six Sigma methodology that will be very beneficial to readers.

There is no other Six Sigma book that talks about the ASQ Certified Six Sigma Black Belt program. The ASQ certification is gaining acceptance and offers an alternative to experienced professionals who want to have their competency level tested and recognized. The ASQ certification has been presented as a supplement to the Black Belt training, instead of direct competition. The Black Belt training providers must look at the ASQ certification as a cost-effective Black Belt certification.

Considering that every implementation of Six Sigma has not been a successful one, the authors look at approaches to maximize benefits of the Six Sigma initiative. Every executive is concerned about the cost of implementing Six Sigma. Addressing that concern, the authors identify how to minimize cost and maximize benefits.

Finally, a chapter on Six Sigma and other quality systems is an excellent comparative analysis to show how Six Sigma can be a part of a corporation's business management system. Six Sigma is a strategic initiative; its alignment with other business processes is essential to minimize redundancy and waste of resources.

The authors of *Six Sigma Deployment* have attempted to address a strategic initiative in a simple and understandable manner. It is easier said than done. The authors have done a good job at it.

Dr. H. James Harrington
CEO
Harrington Institute, Inc.
Los Gatos, California

PREFACE

Is your competition breathing hotly down your neck? Have you tried other quality-improvement programs without gaining significant ground? Do you sense that change is necessary to obtain your desired business results? This book provides you with a source for understanding the Six Sigma deployment methodologies and strategies. You will also gain a clear comprehension of how your organization can cash in on the benefits of these methodologies. This book is for those of you who are seeking to maintain loyal customers, sustain improvements, obtain world-class performance, or improve your company's bottom-line profitability. Hopefully you are after all of these.

There has been much written about the Six Sigma statistics and technical tools. However, little has been written to address just how one might go about implementing a Six Sigma deployment strategy. This is not a book about tools. It is for all leaders or members of management who want to learn what they need to do to make Six Sigma work in their organization.

Many in business today are feeling the pressure of trying to cultivate loyal customers while simultaneously attracting new customers. Trying to reduce costs and improve processes has been a work in progress for many over the past decades. Six Sigma deployment methodologies provide you with the implementation of various tools and techniques as well as the philosophy and measurements to realize a sustained business advantage. It is important to realize that these methodologies can improve your bottom-line profitability.

The psychology behind any desired change is a process beginning with cognitive recognition that a problem exists. Deployment of Six Sigma methodologies first requires an understanding that you want or need to change in order to remain viable and profitable in globally competitive markets. Recognition is just the first step; actions are needed if one intends to solve or improve any problem area. This book outlines the processes and steps necessary for implementing Six Sigma deployment methodologies in a way that allows you to make these improvements.

In this age of standards, many businesses are implementing various quality management systems. The question of using Six Sigma with other quality systems is addressed in this book to a minor extent. The purpose is to demonstrate how Six Sigma methodologies complement the implementation of quality standards. Six Sigma is an excellent tool to integrate into the quality management system, in order to maximize return on investment.

We have included an appendix on the history of Six Sigma, from its origins to the present. The history is a chronology of events based on research, successes, problems, and quality in general. Also discussed is how Six Sigma is changing the view of quality.

We present our view on the various "belts" (Black Belts, Green Belts) in terms of curriculum and resources. Training is needed in most organizations. While there are many approaches to the training, our intent is to help readers obtain the biggest bang for their training buck. Even so, the investment is not trivial. The real issue is not the investment, but the return on that investment.

No company can become successful in implementing Six Sigma methodologies without integrating the supply base as part of the focus on meeting the needs and wants of the paying customer.

Six Sigma methodologies are more than a process of quality improvement or a set of project improvement tools. When institutionalized, Six Sigma becomes a part of a business strategy that must be planned, executed, monitored, and steered toward the achievement of the organization's desired business results. The executive management of the deployment organization must nurture the Six Sigma business strategies. We provide practical information that will assist you in facilitating a successful implementation of Six Sigma methodologies.

One major key to obtaining successful implementation of Six Sigma methodologies is the alignment of the organization's visions, values, and systems. Forging these into strategic objectives, goals, and plans creates a vigilant focus

If you have not read the works of these men recently, I hope you will make the time to do so again. Every time I read their books or review a class I find a new appreciation for these great men.

As part of my corporate life, I was most fortunate to have instruction from such world-class thinkers and practitioners as George Box, Doug Montgomery, Stewart Hunter, and Don Wheeler. It has been their delivery of training with unbridled clarity and the personal coaching afforded me by these leaders that have contributed to my personal development and intellectual pursuit of statistics and quality improvement processes, methodologies, and techniques. Perry Regier has spent countless hours helping me with problems and specific applications regarding statistical improvement methodologies and techniques. The time and patience Perry invested in me can never be fully repaid. Nothing can ever replace the real world experience of working with real people, real problems, and hands-on applications. Perry continues to be a valued and reliable source of statistical practices.

Co-author, valued friend, and confidant Charlie Wilson not only contributed large amounts of original content to the writing of this book but also served as the coordinator for all that is contained. He has toiled relentlessly to assure that what we have learned through good and bad experiences dealing with real issues is reflected. No one is more willing than Charlie to leap into the fray. Charlie has the keen insight into how people work that can only come from his study and application as a psychologist. Charlie developed all of the concepts and the first draft for all of the artwork included.

My thanks to Praveen Gupta in getting us started. We developed the first draft outline while waiting for flights. When Praveen made contact with the publisher and submitted our outline, he initiated the actions required to translate our intentions into a real effort to publish our collective work we now call *Six Sigma Deployment*.

Our publisher, Elsevier Science, had the patience of Job in helping get this book to market. The fine folks there have provided this co-author team with a constancy of purpose and direction. Through their consistent leadership and positive regard for the concerns of all stakeholders, they have helped to build our tapestry of intentions into a final declarative work that gets our message across to the reader. For all of the many people with Elsevier Science who have helped us—thank you.

Cary W. Adams

There is no project that can be completed entirely by one person. A book like *Six Sigma Deployment* can certainly not be a one-man project. My journey with Six Sigma started in 1985 when Motorola SPS Director of Quality, Jim Eachus, asked me to explore methods beyond the conventional three sigma methods for process control. In 1986, I was fortunate enough to join the late Bill Smith, inventor of Six Sigma, and to guide the implementation of Six Sigma at the first four projects, called Small Wins to Six Sigma. Thanks are due to my then supervisor, Chaitan Daiya, and to the late Bill Smith, who was always cheerful and thoughtful. Moving on in my Six Sigma journey, I came to know Bill Grundstrom, Training Manager at Motorola University, who prepared me to teach Six Sigma courses worldwide for the next ten years.

Outside Motorola, my first success in driving 90 percent improvement per year for two years was realized with Frank Brletich, who is still a pleasure to work with, and now happens to be my partner in the Six Sigma journey. There are many more customers who have contributed to my learning and to clarifying my thoughts about Six Sigma. My sincere thanks go to various editors who have published my articles on Six Sigma in their magazines. The list includes Steve Gold of *Circuitree*, Mike Buetow of *PCFAB*, Lisa Hamburg of *Circuits Assembly*, and Sue Daniels of *Quality Progress*. My first success story on Six Sigma was published in 1990 in *Quality Engineering* magazine, thanks to its editorial staff.

Besides publishers, I would like to thank the coauthors of some of my work in Six Sigma. They include Rajeev Jain, Actoras Consulting Group, for the chapter "Enhancing Six Sigma to Deliver Highest Performance at Lowest Cost," Frank Brletich for coauthoring the chapter "Six Sigma Methodology—The Benefits of a Strategic Approach," and Don Cochrane for contributing ideas for the chapter on Six Sigma and Quality Systems. Coauthoring a book is a learning experience in itself. I could not have found better coauthors than Cary Adams and Charlie Wilson, who have put in relentless effort to contribute towards various sections of this book. My special thanks to Charlie Wilson for the creativity of the beautiful graphics in this book.

Of course, no book can be completed without a publisher. *Six Sigma Deployment* could not have been realized without the support of Michael Forster and the editorial staff at Elsevier Science. Special thanks to Kelly Zerveskes for her commitment to the book, for handling the idiosyncrasies of the authoring team, and for her support throughout the development of the manuscript.

Success is not a one-man show, and it would be unfair to close my acknowledgments without mentioning my colleagues at Quality Technology Company.

Thanks to Rajiv Varshney for research on success stories, and to Kam Gupta and Jit Lodd for reviewing the contents and making suggestions for improvements. Six Sigma is at work!

Finally, I could not have completed *Six Sigma Deployment* without the support of my wife Archana, and my children Krishna and Avanti, who have been inspiring me to do something worthwhile for the last twenty years.

Praveen Gupta
ASQ Fellow

Each co-author brought different strengths, paradigms, writing styles, and development experiences to the table. We spent a fair amount of time melding our experiences into a collective work, which certainly was not without unique challenges. The diversity of our backgrounds carried "baggage" that was not often easy to put aside, even for the greater good of our project. Learning to overcome these challenges strengthened our collaborative resolve to serve the reader with the best usable work we could produce. Coming together as a team has been as great a learning experience, as has the publication process. Thanks to my co-authors, Praveen and Cary, I have expanded my personal and professional growth, and will forever be grateful for having had each of you to call my friend. Kelly, Michael, and Kevin, our publisher team members at Elsevier Science, I personally thank each of you for providing us with professional guidance and support in producing *Six Sigma Deployment*.

Thanks to the following persons who gave of themselves, so that I might gain professional, practical, and "real world" experience in process improvement methodologies that prepared me to contribute to this body of work.

Dr. Ken Pederson, Dave Swisher, and Robert Rogers provided stewardship during my journey into the industrial application of statistical and quality improvement processes. Douglas Carlisle introduced me to the world of industrial environmental improvement processes. Jim Swindel enlightened me on the needs people and procedures development required for advancing a newly constructed industrial facility successfully into commercial operations. Donald L. Verbick, President, Delve Energy, provided me with valuable opportunities assisting management with building and implementing strategic plans. Dr. Dave Larsen, President of Applied Education, got me hooked on the idea of adding elements of "behavior improvement" to quality processes. Mike Brady, former Operations Manager of CogenAmerica, shared the process and employee improvement needs of multi-unit power generating facilities spread across the nation. Barbara King, former ARCO Pipeline EH&S Training

Specialist, educated me on the processes of the pipeline industry and added vastly to ideas regarding coaching skills. Steve Rice, Manufacturing Manager at Deltak, shared ideas for the development of international competitive business advantages. Robert Orr, Senior Consultant Specialist for ISO Plus, added to my knowledge of "other" styles of quality improvement systems. Ronald and Dora Todd, Gastian's Seafood, provided information with respect to how the seafood industries use process maps. Ken and Janice Green, owners of Captain Ken's, provided indoctrination into the processes and procedures of the retail restaurant business. Mike Blanchard, CEO of Blanchard Insurance Super Store, shared his vast knowledge and wisdom on the topics of business strategies, employee development, and economic processes.

Thanks also go to long-time friends Robert and Linda Whitt, owners of Front Row Seats Sports Club. They shared a recent two-year strategic plan that called for a total property renovation and the restructuring of business objectives that would allow them to remain viable in a slim local market niche. Linda and Robert were gracious for sharing details of the successes and improvement opportunities of this massive and costly business venture.

Thanks to long-time friends Howard and Norma Louvier, owners of an array of diverse business concerns that include Houston Valve and Testing Service. Howard contributed to my practical application with "hands on" experiences developing new retail business in the "real world." These experiences have added vastly to my practical knowledge about customer critical criteria.

Thanks to long-time friend Ronald H. Farquharson, CEO of Haden Preston Systems, for introducing product and service improvement ideas relating to the medical appliance industry as it regards international customer critical criteria and development of global competitive business advantages. Thanks to my son, Charles E. Wilson III, IT Manager, Reliant Energy Trading, for his never-wavering support, friendship, and sharing of ideas for this book. Thanks to my daughter Kelly Wilson King, schoolteacher, for her never-wavering support and sharing ideas for this book. Thanks to Mike and Tammye Bridgeman for their ideas and support. Thanks to Mark and Sarah Barry, CEOs of Blue Knight Consulting, for providing aeronautical and computer technology industry business process mapping and improvement strategies. A special thanks goes to my loving wife and partner Sena L. Wilson. Sena contributed intellect, support, and encouragement during our long collective efforts developing this book.

I thank my business partner, co-author, valued friend, and confidant Cary W. Adams. Cary served as our team leader, cheerleader, and motivator during the development of this work. His practical use of Six Sigma methods, tools, and techniques served as our foundation. His leadership kept this co-author team vigilantly focused on efforts toward publishing practical and usable implementation methodologies for Six Sigma deployments that make a meaningful difference. We hope you enjoy reading *Six Sigma Deployment*, and that you will use and share the power of knowledge.

Dr. Charles E. Wilson, Jr.

INTRODUCTION

From the beginning of the quality revolution, Dr. W. Edwards Deming focused on the need for management leadership in the transformation of organizations into world class performers. In *Six Sigma Deployment*, the authors are not teaching statistical tools, but rather focusing on the critical role of management in developing and implementing a Six Sigma strategy for improvement. This book acknowledges that the Six Sigma DMAIC (define, measure, analyze, improve, and control) process is implemented through both the people and the systems and processes that are at work in the organization.

The authors' premise for the role of management is on the mark. If implementation of Six Sigma is to bring about world class performance in a company, then management must be able to see the opportunities for improvement and identify the criteria for prioritizing these opportunities. In order to achieve improvement in these opportunities, management must also be able to select and provide the proper training, career path, and compensation for the people who will lead the projects as Black Belts. Then management must be relentless in supporting Six Sigma and promoting awareness of the need to improve through the use of data in making decisions.

If the reader believes that defects cost money, that fewer defects means lower cost, that lowest costs give the ability to compete, that anything less than ideal is an opportunity for improvement, and that gaining data-generated knowledge for improvement of the processes is the most efficient path for lasting improvements, then the reader will find *Six Sigma Deployment* profitable.

Dr. W. Edwards Deming taught that simply working hard was not enough for world class improvement. He taught that everybody gains when optimization takes place. He taught that there is no substitute for knowledge in making decisions for improvement. He taught that change requires transformation, and that the transformation must be led by people who have acquired knowledge. I think Dr. Deming would have liked *Six Sigma Deployment.*

Daniel W. Baugh, Jr., Ph.D.
Six Sigma Master Black Belt
The Dow Chemical Company, retired

1

WHO NEEDS IT?

Are Six Sigma methodologies designed for a small company or for a large company? Do we need a large reserve of money to implement Six Sigma? Am I ready to implement Six Sigma in my department? Where do I start? What do I need? What help is available? Even more important, why should I implement Six Sigma? Answer all these questions when considering Six Sigma for your company.

These questions make sense from both an investment perspective and a business needs perspective. In tough economic times, the focus must be on improving processes and profitability. Six Sigma is a wonderful tool for realizing dramatic improvement. However, in good economic conditions, Six Sigma is also very important because it helps to improve processes through employee participation and is an excellent tool for improving employee morale. Continuing to improve profits is a never-ending responsibility of leadership, no matter what the economic climate.

Why would a company pursue Six Sigma? Studies show that companies that have implemented Six Sigma have improved their processes by as much as 100 percent. Traditionally, people seem satisfied with 99 percent performance. However, 99 percent performance could mean 20,000 lost articles of mail per hour, two short or long landings at major airports each day, or 200,000 wrong drug prescriptions each year.

Six Sigma implies 3.4 defects per million opportunities, which is almost perfect work. Jack Welch, retired CEO of General Electric, has said that such a quality drive requires the passionate commitment of all employees to achieve dramatic results. Improvement in quality increases employee and customer satisfaction, improves profitability and enhances reputation. Welch has shown the highest level of commitment in implementing Six Sigma successfully at GE.

A profitability equation has two elements: cost of goods sold and price of goods sold. Capacity, quality, demand, and reputation determine the price of goods. The cost of goods includes cost of material, processes, equipment, and people. Typically, excessive cost includes factors such as inefficiencies, rejects, rework, scrap, returns, process inefficiencies, maintenance, and other support operations. The reduction in cost of goods sold improves the margin on sales, so look for opportunities in the cost of goods. Opportunities manifest themselves in the form of wasted resources such as time, equipment, material, facilities, and maintenance.

Based on the scope of savings, one can decide how much investment to make in order to realize these savings. Even specific criteria may be set up for evaluating opportunities and for resource planning. Depending on the complexity of the problem or opportunity, management assigns qualified Six Sigma Green Belt or Black Belt resources to the project.

One can develop a process flow diagram and establish defect level, cycle time, and opportunities. Sometimes, one needs to review the fundamentals of the product design, since many problems are design-related. For the best results, design for Six Sigma utilizes methodologies such as quality function deployment, failure mode and effect analysis (FMEA), risk mitigation, design for assembly, and design for manufacturability. The purpose of this design is to allow the process to produce products and/or services without defect. The customer—not the provider of the service or the manufacturer of the product—defines defects.

Understand that opportunities do exist for dramatic improvement in any business. Management must recognize low-hanging fruit and take advantage of it. This low-hanging fruit can often be harvested without assigning Six Sigma projects. Chronic problems that have a major impact on the process represent great opportunities for great results. View Six Sigma as a serious commitment of resources to achieve serious results. Six Sigma training alone will not provide the best results; rather, a commitment to improve using Six Sigma training will produce the desired profitability. The benefits come only after change.

Six Sigma Implementation

If you want to implement Six Sigma in your company, where do you start? What is your first step? Many executives worry about a false start, failure, or the cost of implementation. We have witnessed companies where executives thought for a long time about implementing Six Sigma initiatives, but each company was unable to act. Those companies no longer exist, and many other companies are currently facing a similar situation. To paraphrase Mark Twain: Even if you are on the right track, if you don't move fast enough you will still get run over.

To implement the Six Sigma initiatives successfully, one should first understand a company's performance. A strategic plan with key objectives, metrics, goals, and plans will help to focus the organization. Six Sigma initiatives focus on improving profitability and achieving sustained strategic objectives, rather than on simply improving quality. However, knowing the cost of poor quality is a starting point. Many executives are surprised to learn the amount of money wasted in a company. Most companies do not even have measurements for tracking the cost of poor quality (COPQ). The components of COPQ are internal failures (scrap, rework, and lost capacity), external failures (field failures, warranty cost, complaints, returned material, and lost business), appraisal (inspection, testing, and audit) and prevention (quality planning, process control, improvement, and training). External failure costs are typically grossly underestimated. Think of recent failures that have essentially destroyed well-known companies.

Understanding waste streams in the system can help identify those areas to attack that have a direct bearing on profitability and margins. Then, executives must plan to eliminate or dramatically reduce waste streams and to recognize competitive advantages. Six Sigma initiatives are for companies that want to be profitable and to be world-class organizations. A typical company, managed by executives looking at averages and satisfactory performance, achieves average results. The average North American company typically experiences three (3) sigma performance. That means half are at less than three sigma!

Once the implementation of Six Sigma is understood as an economically viable methodology to achieve the organization's strategy, management must consider giving it the highest priority. Clearly identify any competing initiative, conflicting priority, or strategic initiative in progress. Besides economic viability, other Six Sigma success factors are:

- Leader's commitment to Six Sigma
- Common language to be used throughout the organization

- Aggressive improvement goals that will force process reengineering continually
- Innovation as the key to achieving dramatic improvement
- Process thinking for decision making based on facts, not emotions
- Communication to maintain the continuity of and interest in Six Sigma
- Metrics for assessing the next steps to achieve dramatic results, because correct measurements are necessary to achieve measurable improvement
- Improvement as a way of life, meaning that the company must plan to improve quality every day
- Employee engagement by making the Six Sigma initiatives a rewarding experience

The most critical success factor is the personal involvement of an organization's chief executive. This may inspire employees to do their best because they feel they are working for the chief executive. Employees are uplifted when they feel that they are involved in achieving a higher cause instead of just performing daily duties.

Uncertainty about initiating Six Sigma comes from several unknowns about the amount of effort involved, level of financial commitment, potential for success, and fear of failure. One way to overcome the initial uncertainties is to establish specific, measurable, attainable, realistic, and tangible goals. Thinking positively about how to achieve these specific goals and how to benefit from using Six Sigma in terms of awareness, cultural change, focus on quality, savings, and profitability will expedite the decision-making process. Decision making must be based on the company's needs to reduce waste, save money, and improve profitability in terms of time, resources, and the financial investment associated with implementing the Six Sigma initiatives. Finally, the executives' decision, managers' motivation, and employees' enthusiasm will ensure the success of the Six Sigma initiatives.

How can you determine if your company can benefit from Six Sigma implementation? Look for these characteristics:

- Quality focus and objectives not clearly defined and communicated
- Executives who think quality has nothing to do with business and profitability
- Lack of measurements (levels and trends) to track operations performance, including reject rate, rolled yield, COPQ, design effectiveness, cycle time, inventory levels, employee skills development, and financial performance

- Lack of measurements that lead to centralized decision making (executives make the decisions)
- Executives who are busy fighting fires, making an effort to "look busy," and badgering employees
- Employees who are afraid of management, are reluctant to take the initiative to improve performance, and feel that no one is listening to their concerns
- A major change in competition within your industry that has occurred or is about to occur

To overcome such bottlenecks to higher profitability, a company must establish focus, develop a business initiative, and implement measurements. All employees must recognize the value of improving profitability, have passion to achieve improved results, and become committed to improving business performance.

After implementing the correct performance measurements, establishing dramatic improvement objectives, and committing necessary resources, executives need to reward and recognize employees' success through profit sharing and promotions. Identifying an area or division for piloting the Six Sigma initiative is a good way to develop a successful program. Having in-house success stories can be a great way to create interest in other departments, divisions, and management sectors.

To implement Six Sigma methodologies, the plan must include training of executives in Six Sigma methodologies. Company executives must understand the concept, steps, requirements, expectations, and management to actively participate and contribute to the success of the project. Employees should never have reason to doubt that the executives' priority is to improve profitability, passion for methodology, and staff contribution. Employees must understand the consequences associated with applying or ignoring Six Sigma methodologies. The entire company—executives and employees—must have a common goal, a common objective, and a common priority; they must see Six Sigma as providing assistance in achieving those important goals.

Executives must select employees to champion various initiatives. These employees must have a passion for improvement, be success oriented, have a positive attitude, and be people of action. Committee approaches do not work; instead, teams where each player plays one's role to its fullest are preferred.

Employees should receive awareness training to ensure successful implementation. Employees selected for additional training to Black Belt or Green Belt levels of competency possess a willingness to apply statistics, and a willingness to learn new tools. These employees are often keen observers, experimenters with a willingness to investigate and to solve problems. Black Belts will eventually become team leaders and project managers to implement Six Sigma methodologies. Green Belts work as team members, under the guidance of Black Belts, to learn advanced skills and to expand the development of in-house Six Sigma expertise. One way to estimate the number of Black Belts needed is to have one for every million in revenue. Another estimate is that 0.5–4.0 percent of the employees should be Black Belts. Train approximately ten employees as Green Belts for each Black Belt. Every manager should be a Green Belt or higher within the first five years of deploying Six Sigma. Six Sigma initiative steps include the following:

- Perform financial analysis to understand profitability, COPQ, and essential contributors.
- Establish profitability objectives.
- Measure contributors to profitability.
- Establish business objectives and define values.
- Recruit a firm for Six Sigma training and implementation guidance.
- Define projects and develop plans to realize improvement.
- Conduct executive, Champion, Black Belt, Green Belt, and Awareness training.
- Solve problems and develop solutions to reduce waste.
- Monitor the progress of projects and provide support as needed.
- Celebrate and publicize successes.
- Learn lessons from small wins and optimize approaches.
- Institutionalize the Six Sigma initiative company-wide.

Six Sigma training is expensive unless implemented correctly. The objective must be to maximize value by achieving the organization's strategic objectives—not to save on the cost of implementation by neglecting the potential for higher profitability. The benefits of implementing Six Sigma are too attractive to overlook additional business, increased revenues, improved operating margins, higher turns, increased earnings, and improved customer relations.

Deployed correctly Six Sigma should pay for all of the associated costs several times over. It is not unusual for one highly successful project to return enough to pay for the entire Six Sigma training and investment.

Unless you have proof through measurements that the cost of defects in your company is less than average (remember, half of all companies are average or less), we are going to assume that your company is average. At three sigma (the North American average company), about 25 percent of gross revenue is consumed by defects. At Six Sigma less than 5 percent of gross revenue is lost to defects. That difference of 20 percent of gross sales has the potential of going to the bottom line as additional profit.

Would you like to have that addition to your profits?

Do you have a plan that promises to move that money from the cost of defects to profits? If your answers are YES and NO, then your organization needs Six Sigma!

2

WHAT IS SIX SIGMA?

WHY SIX SIGMA AND NOT FIVE OR SEVEN SIGMA?

Based on world-class process performance and statistical analysis of real-world processes, the Six Sigma quality levels relate better to customer expectations. Five sigma will not meet customer requirements, and seven will not add significant value. Six Sigma's 3.4 parts per million is close to perfection, and that makes it a more attainable and realistic goal to achieve.

WHY IS SIX SIGMA IMPORTANT?

The process aspect of the Six Sigma methodologies is fundamental to any process in a business. If a business is not practicing the Six Sigma methodologies as a process, then they probably will experience disconnects between their Customer's Critical Criteria, their expectations, and the delivery of quality products or services. Six Sigma deployments focus on the customer expectations, and concentrate on the best applications of scarce resources within the organization to those processes that substantially contribute to the bottom-line profitability of the organization.

HOW CAN YOU USE SIX SIGMA IN YOUR ORGANIZATION?

First, one must understand the principles of Six Sigma methodologies before attempting deployment. Occasionally, people jump to implementation and training of Black Belts quickly, in an attempt to gain a specific advantage.

This frequently leads to chaos and confusion. The most common problem experienced in jumping ahead of the game is that we measure the wrong processes. Meaningless measurements cost us more, waste our valuable time, and create confusion as to our strategic goals.

Does Six Sigma Apply to Service Operations As Well As Manufacturing?

Six Sigma is a process and product improvement methodology. All organizations, whether producing products or providing services, are a collection of processes. Therefore, one can apply Six Sigma methodologies to a process that provides products or services. Manufacturing operations have tangibles, unlike service processes, which do not produce widgets. Service operations have more people involvement but may not have complex technological issues. Each process has its own set of advantages and problems. Six Sigma requires one to think in terms of processes.

What Can I Do to Implement Six Sigma at My Company Successfully?

Six Sigma methodologies are not silver bullets. In order to implement Six Sigma, one must acquire specific education to comprehend the philosophy, principles, measurements, and methodologies as process requirements. Next, you must conclude that you have opportunity, as it exists in your organization, and then decide if you are willing to invest the time, people, and effort to realize the desired gains. The organization's executive management team must nurture the Six Sigma deployment processes. This team must not waiver in their commitment and communications efforts regarding the Six Sigma deployment processes. Six Sigma is not a spectator sport; the management team must lead.

Why Do Some People Criticize Six Sigma?

There is a lot of misunderstanding about the principles of Six Sigma. Some people believe Six Sigma means just the measurements. It is complicated, at best, to realize the organization's desired benefits, if you manipulate your measurement numbers. Six Sigma deployment provides a structured and proven process that assists in defining, analyzing, measuring, enhancing, and controlling sustained improvements aligned to the strategies of the organization.

Occasionally, people state that 3.4 parts per million is not a realistic goal. This misunderstanding can exist because these people have not understood the methodology of Six Sigma or experienced significant competition within their market niche. When one commits to be the best, any goal becomes a reality through innovation and hard work. However, the Six Sigma training by itself will not put forth anything meaningful for a company. One must plan to use Six Sigma as a process in order to realize its benefits, just as you use a tool in your toolbox. Remember, tools are designed to serve you, not to be served by you. Any new idea receives some criticism initially, due to the lack of understanding, and that is often healthy. For some, Six Sigma deployments are a paradigm shift. Any change of the status quo creates a fear of loss, of going back to zero, anxiety about losing personal power or position. People are at the heart of any change or improvement process. It is for this reason that we recommend adding understanding of self and others to the Six Sigma statistical training. Gaining an understanding of why we do what we do can provide greater skills of versatility. Versatility skills can assist in reducing interpersonal relationship tensions and improving our productivity. It is very important to have some understanding of the basic behaviors of ourselves and others so that we may be better equipped to reduce tensions and improve interpersonal productivity.

WHAT IS THE MOST IMPORTANT COMMODITY IN THE ACHIEVEMENT OF SIX SIGMA?

Six Sigma deployment, integrated into the company's strategic plan, can facilitate the achievement of your strategic plans in a faster and more efficient manner. Several companies have implemented Six Sigma with envied success. The success or the failure of any initiative is largely dependent on the degree of commitment and leadership from the executive management team. This is especially true for any successful Six Sigma deployment initiatives.

Six Sigma has at least three different meanings, depending upon the context. The first meaning of Six Sigma is a management philosophy. Six Sigma philosophies are a customer-based approach used to recognize that product and service defects are expensive. Six Sigma philosophies claim that the lowest cost, highest value producer is the most competitive provider of goods and services. Six Sigma is a way to achieve strategic business results. Another answer is that Six Sigma is a statistic. Six Sigma processes will produce less than 3.4 defects or mistakes per million opportunities. Many successful Six Sigma projects do not achieve a 3.4 parts per million opportunities defect rate, or less, but this simply indicates that there is still opportunity. A third meaning of Six Sigma is a process. To implement the Six Sigma management philosophy and achieve the Six Sigma level of 3.4 defects per million opportunities or less there is a process that is used. The Six Sigma processes are: define, measure, analyze, improve, and control.

When answering the question "What is Six Sigma?" one must understand that Six Sigma is not a set of new or unknown tools. Six Sigma tools and techniques are found in many improvement methodologies, such as total quality management. Six Sigma is the application of the statistical tools and techniques to selected important projects at the appropriate time and in alignment with the organization's strategic plans. We believe Six Sigma is the structured application of the tools and techniques applied on a project basis to achieve sustained strategic results.

When institutionalized, Six Sigma is part of a proactive business strategy that is planned, executed, monitored, steered towards success, and nurtured by the executive management of the deployment organization. Are you able to correctly define or explain your organizational vision? Has the vision become just so many words on paper that you see and accept with comfort as just another part of doing business? Does your business executive management team expect these words to be turned into actions and behaviors by each of the organization stakeholders? Is the vision a living strategy intended for all the business stakeholders to follow? Do you have demonstrated agreement among the leadership of the organization on what that vision means? Does the entire organization know, understand, and believe in the vision? Are you unsure of the answers to these questions? Try this test at your next executive leadership meeting. Ask everyone to write out the vision of the organization, and do not allow people to look it up or to read it off a poster on the wall. Keep each written response anonymous. When the participants have completed the assignment, collect the statements and read each written vision aloud. Ideally, the vision statement variation will not be one of substance, but do not be surprised if it is.

Have your stakeholders become adapted to the stimulation of the organizational vision? When changes do occur, will the stakeholders possess the wherewithal to measure against that standard of the organizational vision? Will they be able to sustain the organization plans, goals, and objectives?

We believe that visions, values, systems, and strategic plans are living foundations for all stakeholders within the organization to understand and follow. The living foundational tools should be the guides and blueprints for all the actions and behaviors of those stakeholders.

Much has been written about Six Sigma methodologies, tools, implementation, and the results obtained by such notables as Motorola, General Electric, and Allied Signal. If your organization is not counted among the Fortune one hundred or five hundred businesses, can you afford to undertake the commitments of Six Sigma methodologies as a business strategy?

Businesses that actualize successes, in any or all of these areas, are more likely to encounter greater sustained profit margins and increased market shares. This in turn results in the business's customers obtaining higher quality and reliability of products or services and receiving more choices. These organizations then become the winners in the game of business.

Profits promote efficiency in our businesses by reducing costs. Efficiency is gained by improving equipment usage, improving management methods, providing appropriate skills training to stakeholders, and by implementing our living strategic plans, goals, and objectives. That said, how can our businesses sustain these improvement results? How can we use improvements as a focused business advantage to make additional progress towards our vision of world-class performance and improved bottom-line profitability? It is easy to say, yet it requires the commitment and dedication of all the organization's stakeholders. Simply stated, world-class performance, the winners' circle in the game of business, is accomplished by developing a strategic plan based on understanding each of our Customer's Critical Criteria. We must map out and understand our business processes from end to end, in order to benchmark success or failure in respect to our Customer's Critical Criteria. Upon analysis of these data, we begin selecting and prioritizing significant Six Sigma projects designed to improve our processes, products, and services. The result is a profitable business advantage in the competition to sustain our Customer's Critical Criteria. We are on the road to winning our game of business.

Remember, the customer is always your boss! Today your business competitors are attempting to cut costs, to add value, and to improve customer service. Profits put economic strength into the hands of the customers. It is a simple fact that customers can purchase from you, or they can purchase from anyone else in the world who offers products or services they seek. So how do you cash in on sustaining loyal customers, and attract new ones? We believe the answer is to implement a structured Six Sigma deployment initiative, not as a bolted-on process, but rather one integrated into your organizational culture.

Defining, measuring, analyzing, improving, and controlling your organization's machinery can improve the efficiency of your business's operations, improve your quality, and improve cycle time.

Defining, measuring, analyzing, improving, and controlling your organization's use of money, or the lack thereof, can affect production needed to build new facilities and to revitalize and/or add new equipment. Money can provide new skill sets needed by your stakeholders, and money can secure new technology. Gaining access to cheap raw materials and related resources can have far-reaching effects on your business's productivity. Innovative, creative, collaborative, and insightful management can significantly improve your business productivity.

Defining, measuring, analyzing, improving, and controlling your business methods can provide the knowledge and skill for analyzing your processes. Method analysis can provide valuable information regarding your performance as compared to each of your Customer's Critical Criteria. Understanding your business methodologies can provide the paradigm changes that lead your business to sustainable improvement and increased profitability.

Defining, measuring, analyzing, improving, and controlling your management methods and styles can lead you to increased productivity and profitability. Managers who have not kept abreast of today's global marketplace changes, competitiveness, and productivity improvement methodologies are likely to lead you to rapid business result failures and downward spiraling profits. Managers must have an understanding of the needs of workplace stakeholders, as well as an understanding of each of their Customer's Critical Criteria. Managers emphasizing short-term gains do so at the expense of the long-term welfare of the business. If the manpower, the workplace stakeholders, have not kept abreast of essential skill sets required for a competitive and profitable business, again we have a recipe for disaster. All said, how do businesses remain competitive and profitable in today's environment? We believe Six Sigma Deployment provides many of the answers. However, it is just another of the tool sets found within the business's arsenal of survival tools.

Six Sigma deployment methodologies can assist a business in discerning the internal factors—things the business performs—that positively affect the ability to remain competitive. Six Sigma methodologies assist in discovering the external factors—things happening outside the business—that influence economic performance. Six Sigma deployments require strong executive management and Champion commitments. Six Sigma must become a business language and culture used by all organizational stakeholders when discussing measurements of the performance of the business's processes, products, or services. Six Sigma demands a significant sustained improvement to reach the required 3.4 defects per million opportunities rate or less. Six Sigma deployment methodologies produce a nearly perfect, always-room-for-improvement process, not a perfect one.

Realizing aggressive goals requires more than minor adjustments in your business processes. Process thinking is reinforced by the Six Sigma methodologies, requiring one to analyze the process of doing work, for example. Establishing Six Sigma as a common goal means universality throughout the organization. It requires training all stakeholders in the methodologies, processes, tools, and techniques. Six Sigma deployments use and act upon measurements throughout the organization. Process and product performance goals are set in terms of sigma for ease of comparison, a sort of apples-to-apples process. The organization's procedures for product and process development must be modified to ensure effective implementation. Integration of Six Sigma methodologies into every aspect of the business—including, but not limited to, sales, purchasing, manufacturing, engineering, research, and supplier partnerships—is vital to the success of this business improvement process. Finally, it is most important to establish significant goals aligned with the strategic plans, to establish measurements, and to expect continuous progress toward the desired business results of 3.4 defect per million opportunities or less. Let us examine the steps to achieving the excellent performance and profitability gains that are expected when using Six Sigma methodologies. In Figure 2.1 everything is built from the Strategy and implemented through People and Systems. At Moments of Truth with Six Sigma performances, everyone in the organization will be moving the organization toward world-class performance.

It does not matter if your organization is a large multinational or serves only a local market. If you do not possess world-class performance, both in products and services, it is unlikely you will continue to prosper. Today's economy is worldwide in its reach, and the competition is worldwide in scope. Products manufactured in India, Canada, Europe, and South America, and then sub-assembled in China and the Philippines, can then be shipped to Canada and

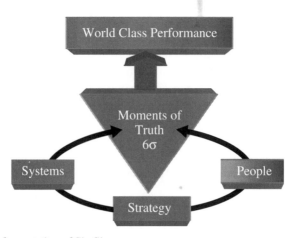

Figure 2.1 Implementation of Six Sigma.

Japan for final assembly and sold in the United States. Protected economies are opening to worldwide trade and finding that the competition is in an entirely different part of the world. Survival in the game of business today directly depends upon your willingness and ability to understand Customer's Critical Criteria, and to produce superior, cost-effective, reliable, and in-demand products and services as compared to others in your particular business.

The drive to world-class performance starts with an organizational strategy. We will have more to say about strategy later. We define strategy as organizational actions in the marketplace that produce competitive advantage. If it does not involve actions in the marketplace that deliver a competitive advantage for the organization, we contend that it is not a strategy.

PEOPLE

Strategies are implemented first through the development of people. This includes skills, teamwork, individual capability, leadership, and social interaction, as well as understanding the personal behavioral comfort zones, quirks, attitudes, and values, and the versatility of people within the organization.

Gaining a basic understanding of personal behaviors and developing the skills of versatility are among the first steps of instilling employee motivation. For the sake of simplicity, we divide people's lives into two major categories, work and personal. At work, people's activities are driven by the accomplishment of the objectives, goals, and plans necessary to achieve the strategic plan of the organization. Most people will have six to nine major work-related goals. Some of these may be short-term and others longer-term. In our personal lives, most would agree that the major components fall into the following categories:

- Spiritual
- Familial
- Social
- Physical
- Financial
- Mental

Now complete a little test using the graphic in Figure 2.2. Consider the center of the circle to be at zero percent satisfaction, and the outside circle as 100 percent satisfaction. Shade in the percentage level of satisfaction you feel that you currently possess, within each of these six areas. There are no "right" answers; the only issue in determining the correct response is the level

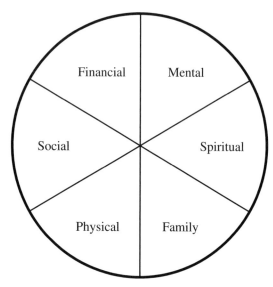

Figure 2.2 Personal life wheel.

of satisfaction you have with your current standing within each of the areas. What might be 95 percent for me in a specific area may be less than 5 percent for you.

Now imagine you removed the outer rim from the wheel. How well does your personal life wheel roll? If it is like our wheel, you will experience some bumps because some areas are much better than others, but none quite 100 percent, and some relatively low.

Now let's consider the other major area of most of our lives, WORK. Figure 2.3 represents another wheel, divided into the different goal areas for work. We have shown six, but you may have a few more, or a few less, depending upon your specific situation. Write in this wheel the goals that you have relative to work. Then, within each goal area, shade in how well you are doing. The hub is again zero percent satisfaction, and the outer circle represents one hundred percent satisfaction.

The question again is: How well are you doing? Equally well in all areas? If your wheel is like ours, again you will have some bumps and dents with the outer rim removed.

For most of us, both of our life wheels, work and personal, have a few bumps and dents that make rolling them through life not as smooth as we might like.

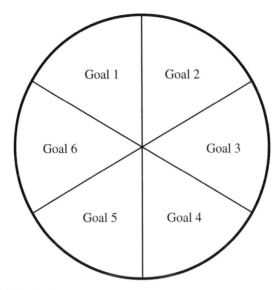

Figure 2.3 Work life wheel.

Figure 2.4 Combined life wheels.

When both wheels are considered at the same time, we see them as a bicycle, as shown in Figure 2.4.

The bicycle clearly shows that the two life wheels are connected. Can you think of specific situations in your life where your personal life was impacted by what happened at work? Similarly, can you think of a time or times when your work was impacted by what was happening in your personal life? Most of us have no problem in thinking of situations for both. We have never found anyone who could not identify occasions when work had influenced his or her personal life, and personal life had influenced work.

If that happens for everyone, then it is naïve to think that we can get people to perform their absolute best in implementing the organizational strategy if the relationship between their work life and personal life is not recognized and taken into account. The first step in this realization is for people to reflect on the relationship of the two life wheels for them as they ride their bicycle through life. Which wheel is driving, and which wheel is steering? It is not our purpose to answer that question for anyone, but to cause you to think about it if you have not before. We also believe that this is not the decision of management or supervision, but belongs to each person on an individual basis.

Accomplishment and progress on work-related goals are legitimate areas of concern for management and supervision. The goal areas should be consistent with achieving the organization's strategic plan in a timely fashion. One of the many things that Six Sigma can do for individuals is to teach a process for defining their goals, measuring current performance, analyzing the situation, making improvements, and then establishing control and monitoring to sustain those improvements. No matter where one is in an organization, the define, measure, analyze, improve, and control steps can lead to the accomplishment of more goals. You do not need to be a Black Belt or Green Belt in order to apply this basic process. Of course, if you understand some of the analytical and problem-solving tools, you will be more efficient.

Because learning these skills will make you more efficient in your work, we think that it only stands to reason that the same skills and learning be transferred to an individual's personal life. We are not proposing that management and supervisors should be tracking personal life goals, but we are suggesting that the process for achieving work-related goals can be transferred to people's personal lives and help them accomplish more of their goals in that area. The training and development of people so they can accomplish more of their personal goals, using what they learn at work, has some very selfish work-related motives. Since the personal life wheel and the work life wheel are part of the same bicycle, if we can help people learn how to do better in their personal life areas, it will make that part of their life "roll" more easily. That translates directly into making the accomplishment of work-related goals easier and moves the organization toward the accomplishment of its strategic objectives.

Any strategic implementation plan that ignores the people in the organization, or only considers them in the context of their work wheel, is missing an opportunity to speed up implementation of the strategic plan and move towards becoming a world-class organization. The trick is to provide the learning process to help people in their personal life, without moving supervisors and managers into the position of treating employees' personal life wheel the same way they do the work wheel. It is a delicate balancing act that requires effort but can pay dividends as both the work and personal life wheels roll more easily and effectively for each employee.

When dealing with people, we find that most will live up to the expectations and assumptions made about them. This is commonly known as the Pygmalion effect. In George Bernard Shaw's play *Pygmalion*, Professor Henry Higgins asserts that he can take a Cockney flower girl and, with some focused training, pass her off as royalty. He succeeds and she fools everyone. However, an essential point lies in a comment made by Eliza Doolittle, the flower girl, to Higgins' friend Colonel Pickering: "You see, really and truly, apart from the things anyone can pick up—the dressing and the proper way of speaking and so on—the difference between a lady and a flower girl is not how she behaves, but how she's treated. I shall always be a flower girl to Professor Higgins, because he always treats me as a flower girl, and always will, but I know I can be a lady to you because you always treat me as a lady, and always will."

In any organization, the people will generally act exactly as they are treated. Those persons expected to take responsibility and act independently will, while those we expect to wait and only do what they are told also will do exactly that. One of the many challenges in developing people is to allow them to act as independently as their abilities allow. Training, development, and opportunity all will make the people side of implementing a strategy more effective.

SYSTEMS AND PROCESSES

The second primary way to accomplish an organization's strategic plan is through the systems and processes that exist within the organization. At the most basic level, this starts with the individual activities. These activities act upon some given set of inputs, tangible or intangible, then produce an output. This is the traditional SIPOC arrangement. Supplier, Input, Process, Output, and Customer are shown in Figure 2.5. Linkage of individual activities forms the Process, and linkages of Processes are the system.

Figure 2.5 Any process.

One thing that frequently slows or stops improvement is considering the process only within the context of the current function, department, organization, or company. The process does not care what organizational boundaries we as human beings have established. The process crosses all of our real and artificial boundaries. The process does not care if it starts in one company and moves through several different departments and then on to another company with multiple departments and finally on to several other companies, or if all of the activities occur within different functions or departments of the same company. Some people have tried to address these issues by vertical integration, thinking that if they just owned the entire process, things would be better. Unless work is complete on the process, pieces can be owned by several different companies or all by one company with no measurable difference in performance. What is important is how the process performs. Unfortunately, poor processes usually defeat good people. The process is a way of communicating to people what is expected of them. Important processes cross the traditional organizational boundaries. Organizational boundaries are usually arranged by function or location. An interesting exercise for anyone considering improvement is to:

- Identify essential customers
- Identify the products and/or services
- Document the processes that deliver the products or services to the customers
- Identify the suppliers to these processes

The following process map, Figure 2.6, illustrates a very simple way to document this kind of effort. Note that time runs continuously from left to right. On the left-hand column are the various functions or departments that are involved with each activity. These can become quite complicated with all sorts of optional paths, and so on. Making the width of the process step correspond to the amount of time required to complete it can provide additional information. A Box Whiskers plot for each step will show even more. This one is quite simple.

The first step in improvement is to understand your starting point. This sort of map can give you some indication. One of the secrets is that the biggest opportunities frequently are between the various steps or between the departments and functions for the same step.

MOMENTS OF TRUTH

We define Moments of Truth as those points where an employee has to make a decision. There is no one to check, no supervisor to give direction, and it

Time

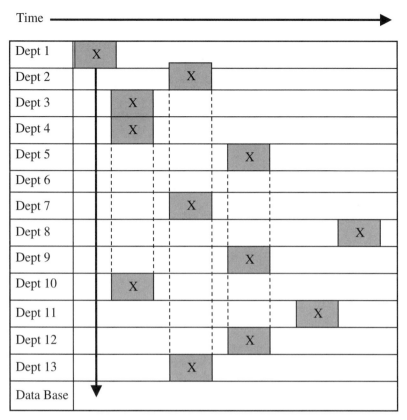

Figure 2.6 Process map.

is highly unlikely that anyone in supervision or management will ever know what the decision was. Our goal for those Moments of Truth is to get every employee to make the same decision that your best and most capable employee would make in that situation, given the same circumstances and information. Maybe on a more personal level, every employee would make the same decision you would make if you were working that job and had the same skill set and information as the employee.

The reason for training and developing people, for improving systems and processes, is to make it easy for employees to make those correct decisions and take the appropriate action. Attitudes, skills, and knowledge must be in place. Systems and processes must be well enough defined to guide employees to the correct action, yet not so restrictive that appropriate judgment and initiative are discouraged. Six Sigma provides the philosophy, measurements, and process to help organizations, teams, and individuals take the correct action when they meet one of the Moments of Truth. A Six Sigma initiative begins with a detailed survey of each Customer's Critical Criteria requirements by senior management. Vocal and visible executive leadership is vital to Six Sigma's success, as to any improvement initiative. These expert leaders then identify activities that are negatively affecting the Customer's Critical Criteria requirements. Project identity and priority, based on this analysis, are established. Management then establishes a cross-functional team to carry out a five-phase methodology: define, measure, analyze, improve, and control. The team identifies the essential internal processes that influence critical customer requirements by taking measurements. They identify the process defects influencing those requirements and collect defect data. Next, the team analyzes the defect data, using standard analytical tools. The objective is to identify opportunities for defects, the variables that cause most of the errors affecting the critical customer requirements.

In the Improvement step, the team quantifies the impact of the variables on the requirements and determines the maximum acceptable ranges—One sigma, two sigma, and so on. Processes within most companies are at the three or four sigma level. They then identify and make the changes necessary to keep process performance within the acceptable ranges. The final step is Control. The team monitors process performance, using statistical process control tools.

Six Sigma projects are a focused approach to meeting strategic goals that directs the efficient application of resources to areas critical to bottom-line success. Long-term, process-focused strategy assists companies in defining and understanding process thinking, making them more proactive as well as productive. Methodologies include the processes and tools associated with defining, measuring, analyzing, improving, and controlling our Six Sigma projects, aligned with our organization's strategic goals and plans. In the next chapter, we will spend time laying out in detail the Six Sigma deployment methodologies and processes.

3

SIX SIGMA DEPLOYMENT OVERVIEW

We established earlier that Six Sigma is a set of business concepts and methodologies designed to address your Customer's Critical Criteria, the demands for high quality, and defect-free product and service processes. Customer satisfaction and continuous improvements are critical factors in the success equation of any business.

This chapter explores the implementation of Six Sigma deployment. We feel that no company can become successful in implementing Six Sigma methodologies without integrating the supply base. Six Sigma has several different meanings, depending upon the context in which it is used. We have discovered that many people attempting to learn about Six Sigma become confused as a result of these many uses of the term and its definition.

Sigma is the eighteenth Greek letter. Capitalized it is Σ, and in lower case σ. The upper case is frequently used as a summation of numbers.

For example:

$$\sum_{i=1}^{n} X_i \quad \text{Sum of all the } X_i \text{ from } i = 1 \text{ to } i = n.$$

The lowercase σ is used in statistics to denote the standard deviation of the population.

The standard deviation σ is the square root of the variances σ^2. The variance for a population is given by:

$$\frac{\sum_{i=1}^{n}(x_i - \bar{x})^2}{n}$$

Stated in words:

Take the difference between each individual value and the mean.
Square that difference.
Sum all the squared differences.
Divide the sum by the total number in the population.

Most of the time we are working with a sample, and the variance (s^2) for the sample is:

$$\frac{\sum_{i=1}^{n}(x_i - \bar{x})^2}{n - 1}$$

Note that the sample variance is calculated using the number of samples minus one. This calculation is due to the uncertainty associated with making an estimate based on a sample.

The convention is that Greek letters are used to describe a population, and Arabic letters are used for samples. It is from the use of lowercase sigma that the phrase Six Sigma draws its meaning as a measure of performance. Each sigma represents a standard deviation. Hence, Six Sigma is six standard deviations. By convention, if not stated, we measure standard deviations from the mean or average in either direction or $\pm$ from the mean.

Two sigma then means the interval from -2 standard deviations from the mean to $+2$ standard deviations, so the total interval is really four standard deviations. Six Sigma is twelve (12) standard deviations. In using statistics, you may be interested in the amount included in the interval or the amount outside the interval.

The sigma levels are translated into the amount outside the interval, if the given number of standard deviations includes the acceptable performance range. So if we say a process is two sigma, that means the amount of variation covered by two standard deviations is all that is acceptable for that measurement. Anything more than two standard deviations from the mean will be outside the acceptable range of performance. The sigma level tells you what the chances are that there will be performance outside the acceptable range (a defect). A 6-sigma process will have 3.4 chances in one million of a defect. A 3-sigma process will have 66,810 chances in one million of a defect. (Later we will discuss why 6 sigma is 3.4 parts per million and not 2 parts per billion.)

The metric or measurement of 3.4 parts per million defects is the first meaning of the term Six Sigma. It offers a standard of performance that can be compared across processes, products, companies, and industries.

PHILOSOPHY

Six Sigma is used philosophically in a more general contextual meaning for driving performance levels to very low defect ranges. The rate of 3.4 parts per million represents a very high level of performance, yet it is not quite perfection. This ideally lets people know when a process performance is in good enough control that efforts probably need to be directed elsewhere.

Low defect rates lead to low costs involved. A caution here: if the process is delivering a high rate of defects and there is an inspection, sorting, and disposition of the failing parts, this is not considered a low defect process. A process that relies on inspection to sort the good from the bad is not considered a Six Sigma process. If one measures the sigma level after the inspection, it is the performance of the inspection and sorting process that is being evaluated. In order to understand the performance of the process, the measurement must occur before the sorting by inspection. The inspection does not have to be performed by human inspectors. The inspection and sorting can use go/no go gages, screens, filters, and the like.

A very low defect process contains less cost involved at essentially every stage. Measurement costs are less, inspection costs are less, and all similar activities that many of us generally accept as a normal part of doing business cost less. There is no rework or repair, no out of specification storage, off grade. The process becomes very simple—the inputs are processed into outputs, which are delivered to the customer. Processes that operate in this fashion avoid all the waste of resources, space, and employees' time, and the other costs associated with producing something that does not meet expectations. Product manufactures of this type will be the low-cost producers. Low-cost producers with very low defect levels tend to capture additional market share.

Anything that fails to meet the Customer's Critical Criteria expectations is considered defective. Defects can occur in the product or service, human interaction, advertising, delivery, accounts receivable, sales, marketing, R&D, and so forth. Everything in the organization has at least one opportunity for a defect, and most have many opportunities. This assumes that there is a reason for the department, function, or activity. Stated simply, processes either do what they are supposed to do, or they do not. Anything that is not what was intended is a defect. Every defect or departure from the desired adds additional cost, which is lost profit or wasted resources.

Most businesses have a general understanding that retaining current customers or getting repeat business is more profitable than finding new customers. Some businesses are likely to have calculated what it costs to generate a sale to a new customer. Even if the exact numbers are unknown, almost all can agree that repeat business is a good deal. Here's something to think about: Who in your business or organization has charge of obtaining new customers? Many companies have large marketing and sales staffs that are dedicated to finding new customers. But who has the responsibility for keeping your business or organization's customers? Some businesses may have account representatives for very large accounts, to assure customers that their needs are being met. When the two are compared, however, the sales budget is frequently larger than the budget allocated amongst various departments for meeting or exceeding the needs of specific customers. The philosophy of Six Sigma is meeting Customer's Critical Criteria, which means driving the defects to a low level and ensuring that there is never a problem with meeting or exceeding customer expectations. Any criteria that remain unfulfilled will lead to customer dissatisfaction.

Customer's Critical Criteria can be as trivial as the tone of voice used by the person answering the telephone. All business or organizational stakeholders who are in the process of delivering to the customer are charged with knowing how they contribute and with ensuring that Customer's Critical Criteria never remain unmet. The ideal level of performance should be at 3.4 parts per million defects or less.

The philosophical approach is another use of the term Six Sigma. Most commonly, it is used to describe the business or organization—for example, "XYZ is a Six Sigma organization." In the philosophical context, Six Sigma does not mean that every product or service of the XYZ organization is at the 3.4 parts per million defects or less level. Rather it means that the philosophy of Six Sigma is accepted in the organization and that there are active project efforts to drive performance levels, in specific Customer's Critical Criteria areas, to very low defect levels. Now the XYZ organization's commitment to improvement is known.

Six Sigma establishes the organization's commitment to strategic plans and goals by first gathering and examining Customer's Critical Criteria regarding defects in the business's processes, products, and/or services as they are related to successfully meeting those criteria. Analysis of information gathered by the organization's intelligence community helps to identify obstacles and defects standing in the way of successfully accomplishing the Customer's Critical Criteria. Before the organization can implement improvements on the defects found, they must create and prioritize project plans.

PROJECT MANAGEMENT

Another use of the phrase Six Sigma concerns the specific steps of project management or problem solving. Define, measure, analyze, improve, and control (DMAIC) are very distinct and important phases of managing a project. Some will suggest that this is the Six Sigma process, but we consider these steps to be the major phases of managing a project. As a process, Six Sigma involves much more than simply defining, measuring, analyzing, improving, and controlling a project.

Six Sigma offers specific tools, techniques, and milestones for each step in DMAIC. Many times, a specific tool or technique can be used in a variety of ways and at different stages of the project. Suggested tools for each major project phase are exactly that, suggestions. The objective is to use the tools and techniques to become more productive and efficient. It is not a requirement that every listed tool be used every time. It is the mark of a professional to select when to use specific tools or techniques. Professional users of the Six Sigma methodologies know when particular tools or techniques are not appropriate. They know if there is a constraint or limit regarding how to use a specific tool or technique to address the issue. We suggest that you make the tools serve you, not that you serve the tools.

The following sections discuss the DMAIC submethodologies of Six Sigma.

Define

The first step of any project is the most important. The Six Sigma methodology of DMAIC is first asking leaders to define our core processes. This methodology is used if existing processes or products do not meet our expectations, and it is the most common approach to Six Sigma project work. It is important to define the selected project scope, expectations, resources, and timelines. The definition step identifies specifically what is part of the project and what is not, and explains the scope of the project. Almost everyone who has ever performed project management activities has had the experience of scope creep. The project scope originally defined is incrementally increased, often with very good reasons, until the project is significantly different than originally stated. Timelines and budgets are affected if the definition is not adequate. Limitations, constraints, specific inclusions, specific exclusions, and other factors are all part of the define phase of a Six Sigma project.

It is important to note the Green Belts activities in the Six Sigma project define stage. Green Belts use the definition tools to identify and validate improvement projects, to assist in illustrating their business processes, to define their Customer's Critical Criteria, and to prepare themselves as effective Six Sigma project team members.

The definition of a project is a cooperative effort between the project manager (Black Belt) and those who have assigned the project (managers, process owners, and Champions). Dedicating extra time and effort during this phase of the project will pay dividends several times over during the life of the project. As an integral part of the define stage, the process or processes involved are documented. Frequently, the first passes at process documentation are at a general level. Additional work is often required to adequately understand and correctly document the processes. As the saying goes, "The devil is in the details."

Six Sigma projects are not a way to put another mouth at the trough of capital investment. Six Sigma projects are expected to deliver the desired returns with essentially no capital investment. This concept should be very clear to all involved in the define stage. Process improvements can be made without spending additional capital. Once the Six Sigma project is completed, you may have identified some areas where investment of additional capital is necessary. Justify capital investment on its own merits. Do not be in a rush to automate or install computer systems for poor processes. This is likely to automate the waste and inefficiency into your system. It is much smarter to improve the process and then add the automation.

Measure

Many think that when they start a journey the most important thing to know is where they are going. While we agree that knowing where you want to go is very important, we believe the first information you need before starting any journey is your current location. The measurement step is asking the Black Belt project manager to quantify and benchmark the improvement opportunity, using actual data. Six Sigma Green Belts deploy specific tools designed to determine critical measures necessary to satisfy the Customer's Critical Criteria. Green Belts also help develop a measurement plan to document process or product performances.

It is surprising the number of people and organizations that start any number of project improvement efforts without having this necessary and basic measurement information. One might keep in mind that a single-point-in-time measure is very unrepresentative of actual overall performance. We need historical data to gain a proper understanding of the actual performance. At a minimum, consider the mean or average performance and some estimate of the dispersion or variation (maybe even calculate the standard deviation). Trends and cycles can also be very revealing. Using two data points and extrapolating to infinity is not a recommended approach. Measurement of performance provides additional information when the performance requirements are understood and competitive benchmarks are established. With measurements in hand, it is possible to make reasonable estimates of the value of a particular project. Alignment with the appropriate key performance indications in the strategic plan becomes obvious to all associated with the project. In today's competitive environment, time is often a most important consideration, and cycle time measurements become a prime metric for improvement.

It is in the measurement stage that every project should be measured for past contribution and potential value in Balanced Scorecard areas. There are many different areas that can be part of a Balanced Scorecard; the classic ones from Kaplan and Norton are identified as:

> Customer Perspective (How do we perceive our customers?)
> Financial Perspective (How do we perceive our shareholders?)
> Internal Business Process (In what processes should we excel to succeed?)
> Learning and Growth Perspective (How will we sustain our ability to change and improve?)

Balanced Scorecard is a concept designed to assist in translating the organization's improvement strategy into measurable actions. Balanced Scorecard provides the organization's executive management and Champion team with a comprehensive representation of business operations and a methodology that facilitates the communication and understanding of business goals and strategies throughout the organization. Balanced Scorecard begins with the organization's executive management and Champion team members creating a vision and strategies. It is from these that the critical success factors are then defined. Measurements are identified to aid goal setting and performance in areas critical to the success of these stated strategies, goals, and objectives. Balanced Scorecard should be considered a performance measurement system, derived from vision and strategy, and reflecting the critical aspects of the business. Balanced Scorecard supports the executive management and Champion team's strategic planning and implementation process.

Creating a Balanced Scorecard for our organization requires that we understand our current position and then identify our vision. A commonly used approach is SWOT (Strengths, Weakness, Opportunity, and Threats). Where is the organization going in the future?

Identifying the strategies is the next step to discover how to get where the organization is destined to go. Next, we define the critical success factors and perspectives by asking what we must do well, regarding each identified perspective, in order to be successful. It is at this point the executive management and Champion team will likely prioritize improvement opportunity projects for hand-off to an appropriate Black Belt. The Six Sigma Black Belt thereafter asks the question, "How do we measure whether all is going as expected?" The Black Belt must consider how to ensure that we are measuring the right effects. Based on this work, the Black Belt project leaders and Green Belt team members create actions, documentation, and reporting plans.

Analyze

Once the project is understood, the baseline performance is documented, and it is verified that there is real opportunity, it is time to do an analysis of the process. In this step, the Six Sigma Black Belt is applying statistical tools to validate root causes of problems. Six Sigma Green Belts will begin the process of analyzing performance data to subsequently refine the opportunity for improvement. Any number of tools and tests can be used. The objective is to understand the process at a level that allows us to formulate options for improvement. We should be able to compare the various options with each other to determine the most promising alternatives. As with many activities, balance must be achieved. Superficial analysis and understanding will lead to unproductive options being selected, forcing recycle through the process to make improvements. At the other extreme is the paralysis of analysis. The ability to strike the appropriate balance is what makes the Six Sigma Black Belt highly valuable. Most Six Sigma projects will have a wealth of opportunity, and it is during the analysis phase that the

Black Belt starts to make decisions about what should be accomplished, and in which order. One of the important considerations is to identify and understand all options or choices, as these will likely contribute to a greater project improvement, considering the efforts and resources available for the project.

Improve

During this phase of the improvement opportunity project, we put ideas and solutions to work. The Six Sigma Black Belt has discovered and validated all known root causes of the existing opportunity. The improvement requires Black Belts to innovatively identify methods to resolve process or product Customer's Critical Criteria defects, to prioritize, and finally to implement these solutions. The Green Belt team assists in creating process and/or product solutions that eliminate the root causes of Customer's Critical Criteria defects. Few ideas or opportunities are so good that they meet with instant success. As part of the implementation there must be checks to ensure that the desired results are being achieved. Some experiments and trials may be required in order to find the best solution. When making trials and experiments, it is important that all project associates understand that these are trials and that they really are part of the optimization improvement effort. The truly elegant solutions deliver benefits at multiple levels simultaneously; each of the Balanced Scorecard areas can deliver significant improvements with a synergy that exists between the different areas. Trade-offs in which one area is improved at the expense of another area are to be avoided. One of the ideas that we want to eliminate in the minds of our organization is captured in this worn bit of humor: fast, cheap, and quality—choose any two. World-class businesses deliver all three. Six Sigma Black Belts are the project managers who develop and implement changes that deliver desired business results that are in alignment with the organization's strategic plans and goals. World-class organizations are creating ways to help their employees make the correct decisions and take the appropriate actions at the Moments of Truth. Moments of Truth are those instances when an employee must make a decision or take some type of action, without the advice of a supervisor, coach, or knowledgeable fellow employee. Ideally, at those Moments of Truth the employee will perform in a manner that is within the organization's high standards and will maintain the Customer's Critical Criteria. When the appropriate systems and people development has been implemented, every employee stakeholder will make the decision and take the action that advances the strategic plan of the organization. When every employee acts or performs in this manner, the organization is indeed on its way to becoming a world-class performer.

Control

Many people believe that the best performance you can ever propagate from a process is at the very beginning. There is an expectation that, over time, things will slowly get a little worse, until finally it is time for another major effort towards improvement. Contrasted with this is the Kaizen approach that seeks to make everything incrementally better on a continuous basis. The sum of all these incremental improvements can be quite large. As part of the project methodology for Six Sigma, performance-tracking mechanisms and measurements should be in place to ensure, at a minimum, that the gains made in the project are not lost over a period of time. The ideal systems and measurements deliver continued improvement. Having an independent financial evaluation of a project implementation at the one-year anniversary is a good way to ensure that improvements are not transitory. In the short term, the Hawthorne Effect can distort understanding.

The Hawthorne Effect is an initial improvement in a process of production caused by the obtrusive observation of that process. The effect was first noticed in the Hawthorne plant of Western Electric. Production increased not as a consequence of actual changes in working conditions introduced by the plant's management, but because management demonstrated interest in such improvements (*Web Dictionary of Cybernetics and Systems*).

As part of the control step, we encourage sharing with others in the organization. When Six Sigma really starts to create phenomenal returns, ideas and projects in one part of the organization are translated in a very rapid fashion to implementation in another part of the organization. The Six Sigma Black Belt must be vigilant regarding the sustainability of solutions used to correct or improve the opportunities, by building process control systems to ensure sustainability.

STRUCTURED APPLICATION

Six Sigma is the structured application of tools and techniques applied on a project basis to achieve sustained strategic results. Key in this definition is structured application.

Six Sigma contains the project management steps of define, measure, analyze, improve, and control (DMAIC). Within each of the major project steps, certain tools and techniques are considered most effective. At each Six Sigma project management step there are specific deliverables that are expected from the Black Belt and project team. We are trying to stack the deck in favor of a successful project every time by providing guidance and instruction about the tools to use and the milestones expected for each step. Because the projects are important to the strategic business results, and we have assigned proficient people to lead them, it only makes sense to do everything possible to ensure a successful outcome. Having consistent and known stages makes it likely that some sharing of information with the appropriate people at each of the major steps of define, measure, analyze, improve, and control will take place. Provide documented reports and updates to the Champion and the appropriate managers. This progress report can be formal or informal, but we strongly counsel that documentation of all actions and decisions be maintained. A storyboard is a nice compact way to summarize this information. The executive management team should consider sharing these reports with the entire organization as a show of commitment and effort to attain the desired results of the strategic plan and the goal of world-class performance.

TOOLS AND TECHNIQUES

There are essentially no new tools or techniques in Six Sigma when compared to total quality management, lean manufacturing, theory of constraints, and other previously popular initiatives. In fact, one of the criticisms of some total quality management practitioners was that they seemed to think that the more tools and techniques applied, the more valuable the project. Six Sigma methodologies use any tool that adds value and improves efficiency or effectiveness. The focus is on achieving results. Training in tools and techniques is important in that their use can help a Black Belt and the project team do a more comprehensive and effective job of achieving desired results. Make the tools work for you, don't work for the tools! This is the mark of a true Six Sigma professional.

Activity-based accounting is a management method of measuring the cost and performance of the organization, according to the activities that the organization performs in producing its output. Activity-based accounting accounts for all fixed and direct costs as variables, without allocating costs based on a customer's unit volume, total days in production, or percentage of indirect costs. This information can provide an integrated and cross-functional view of the organization, including activities and business processes. This methodology tracks the flow of activities in an organization by identifying causal links between the activity or resource consumption and the cost object. The activity flow is characterized through five core areas:

- Resources
- Resource Drivers

- Activities
- Activity Drivers
- Cost Objects (products, services, customers, market areas)

The activity-based costing discipline focuses on effective and efficient management of the activities as the route to continuously improving value received by customers and the profit received by providing this value.

Activity-based management methodology utilizes the cost information gathered through activity-based costing to determine what drives the activities of the organization and to discern how these activities may be improved to increase profitability. The benefits of these two methodologies or tools are the improved effectiveness and efficiency of the organization's processes. These tools can eliminate the non-value-adding work, and improve specific processes by managing the activities that caused the incurred costs. These tools arm the organization's executive management and Champion team with the ability to make effective decisions regarding product lines, market segments, and certainly customer relationships. These tools measure the organization's performance, efficiency, and quality. They increase the value customers receive from consuming products or services. They measure customer and product profitability, and increase and sustain the organization's profitability. These tools provide the organization with real and correct costs.

PROJECTS

Assign Black Belts to meaningful projects as a full-time assignment for an agreed time frame, generally two to three years. Considering senior management's commitment, strategic plans, desired financials, and other Balanced Scorecard expectations, assigning certified Six Sigma Black Belts as full-time project managers ensures the vigilant focus of the organization's resources on successful implementation and achievement of those desired results. There is a place within the organizational structure for smaller incremental improvements, and we encourage you to exercise this option at all levels in an organization. These small-return efforts, however, do not justify the full-time investment of the time and effort of highly skilled and motivated Black Belts. Six Sigma Black Belt projects are developed and prioritized by senior management. They contain significant potential, they are of extreme importance, and they will contribute to the achievement of the organization's overall strategic plan when successfully completed.

SUSTAINED STRATEGIC RESULTS

Many efforts at total quality management (TQM) and other popular initiatives failed to deliver the promised improvements because there was no focus or business reason driving many of the activities. With no benefits, the only thing many TQM efforts did was increase costs and hurt the bottom line. My favorite question for organizations with a failed TQM effort is, "How many had a Re-stripe the Parking Lot project?" It is not unusual for everyone to admit to having had a team assigned to something similar as part of their TQM efforts. The question then is, "What did a Re-stripe the Parking Lot project add to achieving the strategic objectives of the organization?" There is nothing wrong with re-striping the parking lot if it needs it, but I doubt it will make any significant impact on the strategic objectives of the organization. As such, it would never pass muster as a potential Six Sigma project. Six Sigma projects are only approved when they will make a real contribution. Balanced Scorecard areas should benefit from a Six Sigma project. At the top of a Balanced Scorecard are financial and customer results. Use 3.4 parts per million defect rates or less as a way of measuring when a process has improved sufficiently. Confirmation that the results are being maintained is one of the few ways that an organization can continue to build on past success. If, as is often the case with many efforts, there is a burst of improvement followed by a slow and deadly decline until the next burst, it is difficult to build a sustainable organization. With Six Sigma, the results of every project are confirmed at the completion of the project and once again at the one-year anniversary, to ensure that gains have not be transitory. Some organizations do audits after years one, three, and five to confirm that gains are being delivered year after year. Without sustained results, the improvements attained do not return to the organization the promise of lasting improvement and impact on the bottom line.

STRATEGIC PLANNING AND GOAL SETTING

Strategic planning and goal setting can move your organization to world-class performance. The organizational leadership brings about strategic planning. Strategic planning is a process that determines the future of an organization. Goal setting determines the resource allocations needed. Having a well-proven process for strategic planning and goal setting makes the effort more efficient. It ensures that a strategic plan will provide a frame that will shape an organization's future.

Well-organized strategic planning and goal setting involve distinct steps. First, understand current conditions both inside and outside an organization.

Some assessments may be appropriate before starting formal strategic planning. Visualize the future as you desire it.

Implementing the Strategic Plan

Address all parts of an organization in your strategic planning. Leaders in each area look to a strategic plan for guidance in goal setting for their part of an organization. This carries over to all members of an organization as they go about their goal-setting efforts. Strategic planning, goal setting, and a goal achievement process provide a frame for all in an organization to align their efforts with the aim of a world-class performance. Adams Associates Consulting, using Six Sigma, specializes in the synergistic combination of strategic planning, leadership, and total quality management to assist clients in achieving their goals. Six Sigma is a planned use of strategy, total quality management, and leadership development. This is a major difference between Six Sigma and a statistical approach or the teaching of total quality management tools. Planning and behavior considerations are the catalysts that allow all other concepts to be a success.

Issues are selected for special attention as Six Sigma projects. Projects with significant importance are assigned to Black Belts as Six Sigma projects. Thus each Six Sigma project is assigned a leader trained in total quality management. These Six Sigma Black Belts' duties include teaching other members of the Six Sigma project team appropriate total quality management philosophy, interfacing with management, coaching leadership skills, teaching total quality management tools, and changing systems to sustain Six Sigma project improvements.

The executive management and Champion team are responsible for the strategic plan, and for selecting and prioritizing potential Six Sigma project areas. Once a Six Sigma project is understood, the Six Sigma Black and Green Belt teams use appropriate tools and techniques to generate alternatives and implement improvements to obtain the desired results of the strategic plan and goals. Six Sigma projects sustain improvements by using control tools. This is the measure, analyze, improve, and control sequence.

The Six Sigma methodologies require effective team building and employee motivation to fit together. Successful organizations are not involved in team building as an end unto itself; effective team building creates increased employee motivation, and organizational results are readily achieved. Effective team building starts with efforts aligned with the organizational strategic plan.

Employee motivation increases when employees are working on real issues for the organization. Six Sigma projects maintain team-building components and employee motivation elements, which are focused on improving well-defined issues.

Enjoyment of work increases team-building effectiveness. Team building is a tool for improving employee motivation. All teams go through very predictable phases. Knowing and understanding these phases gives leadership the confidence to continue the team-building process even when employee motivation appears to be declining.

Personal self-assessment tools aimed at understanding yourself and others can speed and solidify the team-building growth process. Understanding one's personal behavioral comfort zones, quirks, and skills of versatility is a first step in developing employee motivation. Gaining an understanding of other stakeholders' motivations allows every team member the opportunity to add to the synergy of the team.

4

OPPORTUNITY

Every organization wants to achieve some level of results. Unfortunately, too often not everyone in the organization has the same results in mind. Having agreement on the desired results tends to focus efforts. Even if team members do agree upon the basic description of the results, there is often disagreement about how to measure the achievement of those results. We suggest that one must first understand the business's desired results. Consider the common result areas in Table 4.1.

Agree on how to measure the desired result area. You might ask if your measurement system is capable of producing numbers that are useful for the intended result area. Do all of the affected people have confidence in this measurement system? Are numbers generated quickly enough to be useful? Does the measurement depend upon the level of the desired result? Generally, measurements over a continuum (for example, percentage completion) are more useful than yes/no (for example, done/not done) type measurements.

Customer Satisfaction Opportunities

We cannot stress enough the importance of businesses gaining knowledge and understanding their Customer's Critical Criteria. There exists an interaction between the desired business results and customers (patients, clients, buyers, and so on). Without the customer, it is impossible for any business to sustain itself. Achieving the desired results is frequently a result of customer

Table 4.1 Common Result Areas

Sales volume	Profit before tax	Defect level
Market share	Patents issued	Scrap
Earnings per share	Safety performance	First pass prime
Repeat business %	Product returns	Cost per unit produced
New customers %	Warranty claims	Debt to equity
Cash flow	Environmental performance	and many others

actions. Any business without a focus on customer satisfaction is at the mercy of the market. Eventually a competitor will satisfy the customer's desires. As a reminder, one might consider that there are several levels of customer satisfaction:

1. Dissatisfied customers are those looking for another provider of products or services they desire.
2. Satisfied customers are those who are open to the next better opportunity.
3. Loyal customers are those who return despite tempting offers by the competition.

Dissatisfied customers are an interesting group. For example, are you aware that for every customer who complains there are at least twenty-five customers who do not? Dissatisfied customers will tell eight to sixteen other potential customers about their dissatisfaction with products or services. Having access to the World Wide Web, some are now telling thousands of persons. Ninety-one percent of all dissatisfied customers never purchase goods or services from the faulting company again. A prompt effort to resolve dissatisfied customers' issues will result in approximately 85 percent of them becoming repeat customers. Depending upon the business, new customer sales may cost 4 to 100 times as much as a sale to an existing customer.

Less research has been completed to determine what it takes to cause a satisfied customer to change providers. Why take a chance on mere satisfaction? We understand that loyal customers do not leave their provider of products and services even for an attractive offer elsewhere. At the very minimum, loyal customers will give the current provider an opportunity to meet or beat another offer they have received. We are suggesting that maintaining loyal customers is an integral part of any business. One of the ways to help obtain loyal customers is by having products and services that are so good that there is very little chance you will not meet the Customer's Critical Criteria.

Of course, one of the difficulties is in understanding the true Customer's Critical Criteria. Even when you have the requirements in advance, the customer can and will change them without notice or excuse. Having a good recovery process for a dissatisfied customer is a necessity.

Several surveys have been performed to determine why customers do not give a particular company repeat business. Customers not returning for repeat business gave the following reasons:

- Moved: 3%
- Other friendships: 5%
- Competition: 9%
- Dissatisfaction: 14%
- Employee attitude: 68%

These surveys indicated that in addition to the technical training and job skill training provided to employees, some effort aimed at customer satisfaction and employee attitude is appropriate. Bear in mind that these employees may not be the people normally thought of as salespeople. For example, they may be managers, supervisors, secretaries, accounts payable, engineers, accountants, designers, machine operators, security, truck drivers, loading dock, and so on. If these folks are not helping to cultivate loyal customers, then we suggest that they are hurting your customer retention opportunities. Sixty-eight percent of customers lost by a business are alienated by the business employees' poor attitude!

You must measure in order to appreciate how you are performing in this area of your employees' attitudes. Data indicate that less than 4 percent of dissatisfied customers ever bother to lodge a complaint. Most customers find it easier and more comfortable to take their business elsewhere. Cultivating the customer relationship is key to achieving your desired business results. A passive system that depends upon your customers to inform you without effort on your part is not likely to yield the information necessary to improve your customer retention metrics.

Competitor Opportunities

Every successful enterprise will have competition of some kind. Some may find a short-lived competitive edge regarding their patents, trade secrets, and other efforts. Even in these situations, there is competition. Substitutions for your product or service do or will exist soon enough. They may not be precisely what you offer, but in the eye of the customer they may be viable replacements.

Your competition is constantly trying to convert your customers over to their side. Your competition may come from as near as the business across the street or from anywhere around the world. For many business products and services, location is no longer a strong consideration or concern. Components for everything from $3.00 calculators to billion-dollar facilities are manufactured in one location, shipped to another location for partial assembly, sent on to another location for final assembly, and then delivered to yet another location for customer use. This is not to say that location is not a consideration for your customers—just be aware that they may be talking to a competitor you know nothing about, located halfway around the world.

In addition to location, your competitor may be offering your customer differentiating value in the areas of service, technology, quality, reliability, consistency, packaging, price, quantity, customization, speed, and any number of other areas. New ideas and approaches are presented every second of every day. Some are successful and some are not.

With Six Sigma, one of the objectives is to understand customers' needs, demands, requirements, desires, wants, whims, and if possible those things that customers don't even know they want just yet. Then, with this understanding firmly in hand, the objective is to meet all of these wants and needs with such low defect or error rate that this becomes a strong differentiating element in the minds of the customers.

Competitors may cooperate in certain areas, but they are prohibited from doing so in other areas, at least in the United States. Common areas of cooperation are set out in the established industry standards. Agreements and work in this area can frequently make the entire industry stronger—although if you have ever been part of a standard-setting effort, you may challenge the assertion that what goes on there is cooperation! There are many other areas of cooperation between competitors, such as joint ventures and limited partnerships, to name just a few.

Your customers, if they are not loyal customers, are open to the competition. Understanding your competition may prevent some very unpleasant surprises. Your competition has most likely gathered intelligence about your business as well as the market in general. Competitive intelligence should be accumulated only through legal and ethical methods. Many businesses share a wealth of information in trade associations, journals, conferences, Web sites, press releases, annual reports, and in a host of other ways.

Competition has a very direct impact upon your customer satisfaction. The change from a loyal or satisfied customer to no customer at all can occur with a speed that seems to approach that of light. Your competition can be responsible for that change when nothing within your organization has changed at all. Witness the change in the retail market over the last several decades. Small general stores that carried everything and were located in downtown areas eventually gave way to the larger department stores. Next came the shopping mall, generally located in the suburbs, with a large department store as anchor for the mall. More recently, specialty stores have emerged.

There are two kinds of specialty stores. One is the "manufacturer outlet store," which is frequently in an outlet mall location. Then there are the stores that look like big boxes and are not in a mall. They tend to have a narrow range of products, such as electronics, furniture, music, office supplies, toys, clothes, shoes, kitchenware, and the like. Of course, there has been a similar change in the catalog business, a change from longtime standards Sears and Roebuck, J.C. Penny, and Montgomery Ward (only J.C. Penny has a catalog, and you have to pay for it) to the raft of specialty catalogs. In the last several years, e-commerce business has taken off. Everyone seems to have a Web site and be able to sell over the Web. Most surveys indicate that e-commerce business is not new volume but rather a displacement of other forms of purchasing. This method of purchasing is starting to take the place of older forms. Books, music, airline and hotel reservations, car rentals, office supplies, drill pipe, computer software, computer hardware, and so on, are but a few examples that are in the news almost daily. Industry after industry has experienced similar changes—steel, auto, shipbuilding, communications, and airlines just to name a few. Competition is constantly changing.

The intent of change in a business situation is to create opportunity for totally new customers or to move existing customers from where they purchase now to a new supplier of products and services. All of this puts your customers at risk as far as you are concerned. If they are dissatisfied, they will actively seek out your competitors, whether they offer exactly the same product or service or a substitution for what you offer. Satisfied customers will be listening to the temptress song of your competition and may move according to any whim or reason. It is only your loyal customers who will normally give you the opportunity to meet or beat the competitor's offering.

The best way to develop as many loyal customers as possible is to recognize your customers' desires, needs, wishes, and wants better than anyone else. Next, you must possess the capability of meeting those desires, needs, wishes, and wants with such frequency that no one else can even come close. One of the many objectives of Six Sigma is to perform at such a high level that there are less than 3.4 chances per million opportunities of failing to meet or exceed those customer expectations. Few if any of your competitors will be able to match that level of performance. Six Sigma deployment offers the security that loyal customers will not be leaving and that the business results you want will be there.

Products, Services, Work Processes, and Distribution Opportunities

Some of the more direct controllable factors involved in establishing loyal customers are your products, services, work processes and distribution. As part of the effort to maintain and build loyal customers, consider each of these based on its impact on the customer. All products or services contain some performance criteria and thus opportunity for improving your business advantage. Every time those performance criteria are not met, the result is an error, defect, mistake, omission, off-spec, non-prime, second, reportable, or whatever terminology is used to describe your product or service. These failures hurt customer loyalty, even if the customer never sees them.

Opportunities exist in the concept and early design phases and are much less costly then than at any other time. Costs of defects go up exponentially as a product moves from concept to design, to prototype, to field trial, to full production, and finally into the hands of the customer. Correct mistakes whenever detected. Six Sigma suggests that your efforts in eliminating the mistakes or problems be at the point responsible for contributing the most defects. Defects and variation existing at the beginning of the process are amplified as they move through the process uncorrected.

Mistake-proofing and Poke Yoke concepts applied to the engineering and design of a product pay unimagined dividends later in the process. Early consideration of ways to make a product fail can result in more robust designs and superior performance in the hands of the consumer, if these failure modes are addressed. Some solutions do exist that add essentially no additional cost or complexity. In fact, reduction of complexity will generally add to your business advantage. Function analysis can help you discover paths to some of these solutions.

Feedback of Reliability Engineering data from similar components or products are a valuable resource. Closing the loop from design to field reliability should be part of every design process. It is obvious that there can be an impact on the customer if the customer suffers with less than expected performance from your product or service. Many occurrences of defects in products or services will create an EX-customer.

earth or out of the minds and actions of people, there is a supply chain that must be attended to. There are a number of philosophies about how to deal with the supplier issue: backward integration, developing as many as possible to make the supply item a commodity, competitive bidding, single source, and supplier partnerships, to name a few. In general, the closer a product or service is to a commodity item by definition the more potential suppliers are available. For example, in the U.S., wheat is generally considered a commodity, and there are thousands of wheat farmers seeking to fill the need. On the other hand, Kashmir wool is not quite as readily available in the U.S., and there are fewer providers. Maybe in India it would be a commodity item.

Web site development is a relatively new field that has quickly become a commodity service. The number of people performing Web site development has grown extremely fast. Yet, it is our understanding that search engines index less than 13 percent of all Web sites. Other service providers in this area develop Web sites by including words and phrases that people actually enter in a search engine, and that drives legitimate traffic to the Web site.

Commodity suppliers are generally found at the bottom of the economic food chain. The price they get for their goods or services is generally out of their control, and they are forced to take what a fairly price-elastic economy offers. We are not suggesting that commodity suppliers cannot be successful and profitable businesses. They are always seeking to be the lowest-cost producer and have little if any price influence. They have to be extremely sensitive to all of the cost factors involved in their system. Usually you will find the successful commodity suppliers seeking to differentiate themselves from everyone else in some fashion. Faster delivery, friendlier people, easier credit terms, better service, consistency (lack of variation) of product, location, branding and brand allegiance, customizing, various packaging, lot sizes, and so on are all ways that commodity products and services seek to move away from that pure commodity market to a more specialized and expensive product or service, but one that offers better value.

Your suppliers have a strong and important impact on your products, services, work processes, and distribution. The old cliché is that "You can't make a silk purse from a sow's ear." It seems obvious that the quality of the raw material will have an impact on the quality of the product. There are many companies that attempt to reduce their cost structure at the point of contact with the supplier and pay little attention to the internal processes and systems that use the raw material. There is a multitude of ways that the supplier relationship can affect customer relationships beyond just the physical quality of the raw material.

With Six Sigma the goal is to meet the customer's expectations better than any one else, and at a minimum to be able to deliver goods or services at a defect rate of less than 3.4 parts per million. There are two basic approaches to this performance. One is to have your process centered as well as possible at the target value. This is the on-target component. Classic Six Sigma allows a 1.5 standard deviation shift to compensate for the fact that few if any processes remain stably centered exactly on the target value. The second way to approach Six Sigma is to reduce the variation in the process to a level that even with a 1.5 standard deviation shift leaves the chances of a defect at less than 3.4 parts per million. The objective is really both approaches: "on target with minimum variation."

Suppliers have the opportunity to influence both of the objectives. They are at the beginning of the chain of process steps for your product or service. Assume that you have five process steps in your organization, all of which are performing at better than a Six Sigma level, let's say 3 ppm (parts per million) defect level. This is 0.999997 good at each process step. For argument's sake, there is no inspection and sorting of the good from the bad, you are using what the supplier furnishes and shipping to your customer. If your supplier is at the same 0.999997 performance level, the defect rate for these six steps (five of yours plus the supplier) is $0.999997 \times 0.999997 \times 0.999997 \times 0.999997 \times 0.999997 \times 0.999997 = 0.999982$, or 18 parts per million defect rate. Realizing that this has failed to meet your objective of less than 3.4 ppm defect rate, you and all in your organization put in a lot of creative and hard work and reduce each of your process steps to 1 ppm defect rate. Your supplier, impressed with a 3 ppm defect rate, neglects to control the process, and their defect rate climbs to 13 ppm. The overall performance after all of your work and effort is now 0.999982 ($0.999999 \times 0.999999 \times 0.999999 \times 0.999999 \times 0.999999 \times 0.999987 = 0.999982$).

That's exactly the same place you started, with 18 parts per million defect rate. I'll leave the math to you to prove that with five process steps at 0.999999, even if the supplier were perfect, the best you could do is 5 parts per million defect rate.

These calculations demonstrate a number of important concepts. One is that suppliers can have a tremendous impact on your products and services. Another is that no matter where in the sequence the defect rate appears high, it will affect overall performance. Each step is a supplier for the next process step. Suppliers are not necessarily external to your organization. Think of all processes as a combination and number of simple component process steps. Deming has suggested a drive to reduce the number of suppliers. The performance level for one supplier is difficult enough to maintain, but as the number of suppliers is increased, the variation increases far beyond that with a single supplier.

There are two sources of variation, that within each supplier and the variation between suppliers. Those of you with some math inclination will quickly realize that the overall process will always perform at a level less than the worst single process step. A simple strategy for improving performance is to eliminate process (reduction in complexity), also done in cycle time reduction. Back in the original example, if we had only four steps plus the supplier, all performing at 0.999997, the end result would be 0.999985, creating a 3 ppm defect improvement rate by eliminating one process step.

Some organizations have started to reduce the number of suppliers in order to reduce some of the variation. These organizations understand that it is not the number of companies that they have as suppliers that is important but rather the number of processes that provide the product or service. These organizations generally qualify a supplier on a process-by-process basis and insist on knowing if there is any change to the supply. This could be a change in machinery, a processing step, personnel, storage conditions, or anything else. It is not that they want to run the supplier's business but that they understand just how little it takes early in a process to cause a major impact at the end. Much like the ripple from a rock thrown in a pond, the diameter of the ripple just keeps getting bigger and bigger.

Understanding process variation will drive you to focus your improvements as far up the supply chain as possible. This is contrary to the intuitive approach of experiencing a defect in a product and looking for the last process step before the defect was discovered. Generally, we then attempt to make the improvements at that process step as a first resolution. We contend that the best approach is to document the various process steps involved and obtain data about the performance of each process step. It will become apparent where the biggest opportunities exist. Cycle time reduction and/or complexity reduction studies often eliminate the defect-producing process step.

Employee Development / Involvement Opportunities

Every business and work process eventually requires that people make a decision to do the right thing. As human beings, we are all created with free will and the capability to make decisions. When employees are not making the correct decisions, no matter how good the process or system, problems will soon develop. Active employee development and involvement will assist you to make more of these organizational decisions beneficial. There are an infinite number of small details that no one except the person actually doing the work can ever know. All of this knowledge is valuable and waiting to be tapped for your organization's benefit. Frequently we make assumptions about employee attitudes and willingness to participate based not on the desires of the employees but on their reaction to the way they are treated by supervision. In many organizations we find employees who are community leaders, serve on church boards, are elected officials, do volunteer work, or have their own businesses. In a variety of other ways, employees often demonstrate a capability above what is used in their work. What could happen to your business if your employees brought the same dedication, effort, and thought to work that they freely give away outside of work? Improvements in productivity of 25 percent to 50 percent have been demonstrated when employers are willing to engage their employees.

The synergy of work processes and system improvements can be amazing. A proven effective way to get involvement is to focus on the cycle time of important work processes. Even with uninspired and hesitant team members, it is common to have 33 percent reductions in cycle time. The importance of cycle time reduction goes much deeper than just being able to perform a process step in less time without adding effort. In an organization dedicated to learning how to improve itself, every cycle is an opportunity to learn and improve. An organization with a 33 percent reduction in cycle time has the advantage of lower costs (time is always money), and the ability to do more with the same or fewer resources. Benefits continue to feed on themselves, and the advantages grow bigger and bigger. Teaching people how to use relatively simple problem solving tools and techniques is the easy part of employee development. Usually after just a little training and experience with one or two work-related problems, the basic tools are mastered well enough for most to start using them on their own. When placed in teams, they are prepared to make use of the many specifics that only they know, to improve products and work processes. If your organization is going to approach Six Sigma performance levels (less than 3.4 parts per million error rate), you will have to get your employees actively involved in using problem-solving tools.

Even the best training and development programs cannot ensure that all employees will get involved. One of the prime jobs of supervision and management is to create the environment and systems for employee involvement from the neck up, not just from the neck down. This is not to say that all will choose to become involved at this level. The obligation is to provide the opportunity and the means. It is then the duty of the employee to take advantage of that opportunity. Most employees want to believe in and trust the organization.

Employee training and development processes should stimulate thinking and encourage employees to make positive changes in behavior, attitude, and habits of thought about work. Frequently the biggest changes in these areas must first occur at the management and supervision levels. Turf protection, arbitrary rules, inflexible systems, capricious authority, poor listening, and reservation of the right to make all decisions diminish the likelihood that employees will contribute even a fraction of their capability. True management skill involves the ability to direct, coach, delegate, and mentor individuals and teams, depending upon the situation and the employee's need. Developing management and supervision with the skill and confidence to behave in this manner is not a trivial task. For this reason, we strongly recommend that the employee development start at the top of the organization with a consistent philosophy and approach, backed up with observable behaviors and actions.

Six Sigma provides significant training, tools, and techniques in the area of personal development. Even when no new technical skills or tools are taught, improvements are often impressive. This is especially true when a coordinated effort starts at the executive level in the organization and moves through the managers, supervisors, and employees working on the same concepts and approach. Six Sigma, for example, requires the executive management team to develop, implement, and actively communicate deployment strategies throughout the organization.

Many organizations spend time, money, and effort to teach new skills to employees who are currently using only a small fraction of the skills developed in past training sessions. In order to cash in on the desired benefits of your training efforts, you must provide all the required antecedents and hold trainees accountable for the consequences of the desired behaviors. Management must provide trainees with proper coaching, goal setting, tools, and opportunity to apply training. Higher returns on your investments are realized when you emphasize how and why employee contributions affect the desired business results. This method generally provides employees with a feeling of commitment verses compliance. They feel valued as owners of the desired business results, which is in turn reflected in their work behaviors. Every activity or job has some level of technical skill that must be mastered in order to perform at an acceptable level. Without these, it is much like trying to drive a screw into a board without a screwdriver. Just a simple tool, a screwdriver, can dramatically improve productivity, as well as employee job satisfaction. Demonstrated knowledge and skills are essential. In some cases employees come to the job with all required skills. More commonly your employees will have a certain base level of competence but still will require additional training and development before they can make a positive contribution to your desired business results. Sometimes it can take years for the contribution to pay back the time value of the investment made in an employee. An obvious improvement would be to reduce the amount of time (cycle time) that it takes for new employees to reach the point of net return.

The attitude that employees present at the workplace can be as important as the actual technical skill level. Most of the time, when we speak of someone having an attitude, it means a poor attitude. When speaking of a positive attitude, we generally use a "good" descriptor. Our experience confirms that poor attitude is one of the more common concerns in the work environment. Actually, it is not the attitude that is the problem, but rather the behavior that results from that attitude. How would you describe a poor attitude? Typical responses are attendance problems, marginal quantity or quality of work, interpersonal problems with coworkers or supervisors, poor communications, lack of cooperation in any activity, and so forth. The list is remarkably similar no matter what the job, company, industry, or part of the world. A fairer approach would be to identify the behaviors said to represent this poor attitude. If the employee and others can see or hear them, the behaviors are said to be measurable. If the behaviors are measurable, they are manageable. If the behaviors are manageable, they can in fact be changed. It is unfair to use labels, generalities, or assumptions regarding another person's performances. Perception is said to be reality to the beholder. In that regard, we are all judges of behavior. Judging another's performances can be unfair if our perceptions have no base in fact. Measurement of identified behaviors provides us with a much clearer picture of the performance. Remember that if you can hear or see behavior, you can measure it. If you can measure behavior, you can manage it. If you can manage behavior, you can change it. For example, if your boss said that you have a poor safety attitude, what would it mean to you? What would you change to improve your attitude? If, however, your boss said that you have a poor safety attitude as demonstrated by your seeming lack of regard for the safety requirement of wearing chemical goggles in designated areas, you would have a measurable behavior to work on.

Almost everyone will make this connection between behavior and attitude. Our study indicates that attitudes tend to drive behavior and are a result of our internal values and beliefs, many of which were imprinted at a very early age. From birth to about age six, we are generally living constantly with our parents, who are teaching us right from wrong, good from bad, and so forth, all of which is being imprinted upon us. From age six to about age sixteen, we are spending more time with our peer group. Now we are learning how we want to talk, walk, and dress, what hair styles to wear, what dress codes to follow, what music to listen to, and the rest. This stage is called modeling. Finally, we spend the remaining time to about age twenty-one in the socialization process. It is at this point that we are refining our belief and value systems. These will become our strongly held paradigms for much of our lives. We have to live with the early messages for the rest of our lives. That means that if we as individuals are going to change our attitudes, we must find ways to overcome that early conditioning.

Fortunately, we can make a conscious choice to add to our values and belief systems. Each of us can make the conscious decision to enlarge our individual inventory of experiences. In the proper environment, individuals can learn to examine their values and beliefs, and choose if they want to make a change. The change is not always easy, but the beginning of change lies in changing the habits of thought, our self-talk. The sequence is that our habits of thought (self-talk) drive our attitudes, and our attitudes drive our behavior. All three will have a certain amount of harmony or agreement. To make a conscious decision to change, we need to change the way we think and change our habits of thought.

Changing someone else's attitude is an impossible task. As we said, we must first identify as specifically as possible the behaviors that are contributing to our perception of another's poor attitude. Instead of discussing an attitude, we would rather talk in terms of behaviors. Measuring either directly or indirectly the specific behaviors of concern will lead to the next step in the change process. That step is to ask if it is legitimate to change the behavior in question. Does it adversely affect business, others, or this person? If the answer is yes to any of these questions, then change is in order. Lasting motivation for change comes from within oneself. Change is difficult at best for the person asked to make the change. Those of us doing the asking generally expect change instantly.

Change management requires us to first identify all antecedents (events or conditions preceding) and consequences (events or conditions following) both the desired behaviors and the undesired behaviors. Measuring both antecedents and consequences in terms of the performer's behaviors will provide clues as to why an individual or team is performing in a certain manner. If you can determine that specific consequences are personal to the individual performer, that they are realized immediately and that they are certain to happen, you have identified the motivators of the performer's behavior.

Long-term motivation and change are personal events. In order to help people learn, one must understand that most learning is based on a few basic inputs. One such input to consider is a significant emotional event. Experiencing injuries, being reprimanded, losing a loved one, being incarcerated, and being fired are among the experiences considered by most to be significant emotional jolts.

Almost all of us can remember where we were and what we were doing during some common major events. As a test, if you are old enough, ask yourself where you were when you first heard that U.S. President John F. Kennedy had been shot? What were you doing when you heard about or saw the TV pictures of the U.S. NASA Challenger explosion? Where were you on September 11, 2001?

Significant emotional events do not require effort on our part to remember. They are events that affect us, and we remember them for most of our lives with no effort or conscious decision to do so. Most of us have experienced a number of unique significant emotional events that are part of us no matter what we do. These types of experiences are almost impossible to predict or create and thus are difficult to use as a method of planned learning.

Vision, Values, and Strategic Direction Opportunities

Your organization's strategy should answer some very basic questions:

- Why do we exist as an organization?
- How do we do what we do?
- Where are we now?
- Where do we want to be?
- How can we get there?
- What would tell us if we arrived?

Everything in the world is created at least twice, once in the mind of the creator and then as the actual product or service. There may be accommodations or changes over time, but at some point someone possessed an idea of what needed development. Trusting your organization to grow and develop in a haphazard fashion is taking a large chance on your business future.

Your organization's vision statement of what the future should hold is essential for your business to control its destiny. Vision statements bring organizational focus and direction to avoid drifting and wandering on the tides of current events. For most, that is not an acceptable alternative, and we prefer to have more control of our organizational destiny. A vision statement describes a clear future state or position that represents the ideal of what we want our organization to become. For most, it is something that can never quite be achieved but is still a worthy endeavor. For many organizations, Six Sigma is part of their vision. Achieving a performance level that has less than 3.4 parts per million defect or error rate may seen like the impossible dream, yet for some it has been achieved. If it has not been achieved for every product or service, they are on a path for even more improvement. One of the biggest hurdles to overcome is the realization that no matter how good your organization is now, it could be better. Every defect or deviation from the ideal represents additional costs and waste. A compelling dissatisfaction with those departures from the ideal and the decision to do something about them are requirements for Six Sigma implementation. Once that decision is made, the Six Sigma tools, techniques, and approaches can be learned and introduced within your organization.

Supporting the vision statement is a set of values for an organization. These may or may not be written down; nonetheless, they exist and most employees firmly understand what they are. When the written and the actual values differ, you can be sure that your employees know where the differences are. The value system will have a fundamental influence on how customers and employees are treated and how employees react to management direction and initiatives. For Six Sigma to succeed, there must be a culture of improvement. If this is not in place, creating it is a major implementation step.

Mission statements describe the purpose of the organization. They state who you are, whom you serve, what products and services you provide, and how you make those products and services available to your customers, clients, or patients. The mission statement tells what the organization was formed to do. Some like to include levels of performance in the mission statement. We contend that this adds complexity and confusion. State what you are about, and let the performance speak to the level or quality.

With values, vision, and mission statements understood, you are ready to develop your strategy and direction. To attempt to develop a strategy before values, vision, and missions are clear, understood, and accepted is a major blunder. Quite simply, strategy is the observable actions in the marketplace that lead to a competitive business advantage. Operative words include observable actions, marketplace, and competitive advantage. If any one of these is missing, you may have some nice sounding words, pretty pictures, and flowery talk, but you will not have a strategy.

Common work for developing a strategy includes a number of assessments both internal and external to the organization. From every assessment, strengths, weakness, opportunities, and threats (SWOT) are evaluated. Once a clear understanding is in place, a specific strategic direction is developed. This should be done at the business level. In large organizations with many business units, subsidiaries, or divisions, each may be allowed to develop its own business strategies. Without some supervision, this can result in divisions of the same company hurting each other. For instance, having Pontiac and Oldsmobile competing in a mutually destructive fashion.

Once a strategic direction is set, some general objectives should be accomplished. From those objectives, goals are developed meeting the SMART criteria: Specific, Measurable, Action-oriented, Realistic, and Time-bound. Each goal should have an action plan(s) to ensure that it is achieved. Frequently, some of the goals from one level become objectives for the next level down in the organization. When this happens it ensures that goal-oriented action plans anywhere in the organization can be traced up the organization to demonstrate that they are in support of the top-level objectives and strategy. When there are gaps in that linkage, it is common to have misunderstandings and efforts that are not aligned, and hence not as efficient or effective as they could be.

Six Sigma can be an integral part of almost any strategic direction. For a strategy involving cost leadership, Six Sigma can be focused to improve internal processes, yields, and productivity, to eliminate complexity, reduce cycle time, and in general help gain or maintain low-cost supplier position for your particular product or service. If your strategy includes offering the lowest price in the market, your costs had better be the lowest. Six Sigma also should be an integral part of any customer loyalty strategy. One of the keys to customer loyalty is providing customers with products and services that meet or exceed their expectations every time. Every transaction and interaction between a company and a potential customer is an opportunity to meet or fail to meet those specific Customer's Critical Criteria expectations. Few systems are good enough to offer the desired level of product or service on a consistent enough basis to keep loyal customers without some constant attention and work. The tremendous benefits from having loyal customers cannot be overstated. Jack Welch has been quoted as saying that only when GE's Six Sigma efforts started focusing on the external customers did they begin to see the benefits.

One of the primary mechanisms for management to fulfill its leadership responsibilities is the development of the vision, mission, and strategic direction for the organization. If Six Sigma is part of the strategy, there are many opportunities to develop leadership at other levels in the organization. Leadership in any organization can always be expanded. With Six Sigma, an objective is to expand the concept of leadership beyond that of organizational position. Frequently people equate leadership with some title, position, or rank. Those who have demonstrated leadership capability frequently have these titles. Unfortunately, there are cases where the title or position has been granted before the capability has been developed. This is an indication of a lapse in leadership by those responsible for placing an unprepared person in a role requiring extensive leadership skills. Too often, very successful technical people are placed in roles with leadership requirements without adequate preparation. The best engineer is made Engineering Manager, or the best operator is made Supervisor. Most of us would not allow an untrained civil engineer to design and build a major highway. Neither would many of us willingly go into surgery with an untrained surgeon performing the procedure. Yet, many of our management and supervisory positions are filled with very competent technical people who possess little preparation for management or supervisory duties. In these cases, there are two options. One is to move along and hope that instinct and a good organizational structure will adequately support this leader. The second option is to begin an accelerated development program intended to develop the skills and knowledge base required for a leadership position.

Six Sigma has had a lot of recent press, and many organizations would like to take advantage of the potential gains but may be unsure as to what course of action is required. While there are several paths that can lead to success, some roads are easier to travel and more direct. We consider Six Sigma as a way to achieve the strategic objectives of an organization.

Six Sigma must be anchored to the organization's strategic objectives; otherwise, the process methodologies become a shotgun approach of improvements resulting from projects that have no synergy or common focus. In an unfocused implementation, projects are selected on an ad hoc basis meeting only minimum criteria, usually some financial return hurdle. We are certainly not against projects that return dollars to the bottom line of the income statement. Every project should do that. The issue is that projects must advance the organization toward its strategic objectives in order to obtain the maximum benefit of the Six Sigma methodologies. In theory, savings in a number of diverse areas will each show up on the bottom line of the income statement, providing an advantage to the organization. In practice, the accounting or finance department spends the time to validate each project. Furthermore, it is difficult to persuade management that anything significant has happened or that a weighty improvement is not due to some other combination of factors.

Focusing project alignment with the strategic direction of the organization allows a number of advantages to accrue. First, since the projects are aligned with the strategic objectives of the organization, there is no problem in maintaining a leadership focus on all projects. As projects begin to influence the strategic objectives of the organization, management has a double reason for maintaining an interest and focus on the Six Sigma efforts. These projects apply to the strategic objectives of the organization, and hence are central areas of interest for management. Second, the Six Sigma projects add to the bottom line

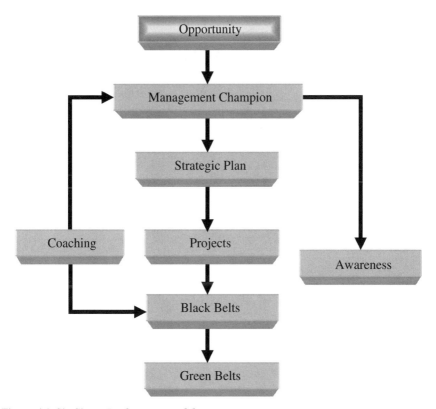

Figure 4.1 Six Sigma Deployment model.

of the organization's income statement. The deployment model we recommend for Six Sigma is shown in Figure 4.1.

Six Sigma Opportunities

The only valid reason for deploying Six Sigma is to take advantage of the opportunities that exist within the organization. If the senior leadership is not convinced that Six Sigma offers the best known way to take advantage of significant opportunities, the commitment and support necessary for successful deployment will be missed.

Some organizations have jumped on the Six Sigma bandwagon, hoping it is a magic bullet that can solve all of their problems with little effort or involvement by the senior leadership. Some organizations indulge Six Sigma processes and methodologies as they would other management fads. This most often results in failures similar to those experienced with other management programs

that are not driven by Customer's Critical Criteria and organizational strategic focus, and are lacking in executive management commitment. Six Sigma
is work, and there is work for every level of the organization. Some of the
most important work is at the executive leadership level. That is where the
foundation is built. If the foundation is weak, it will not take long for problems to appear. Take the time to build a solid foundation for your Six Sigma
deployment.

Some organizations may have been forced into some level of Six Sigma efforts
by a major supplier, or more likely, by customers demanding Six Sigma implementation in order for the organization to remain on the approved vendor list.
This generally results in justification or compliance rather than a commitment.
Few efforts started and/or maintained using compliance as a justification ever
achieve the level of success or improvement promised. The difference between
going through the motions, establishing the semblance of a Six Sigma effort,
and actually deploying it correctly can be several orders of magnitude. The
following is a simple set of calculations to frame the size of the opportunity
that exists in an organization.

Opportunity Calculation

Gross sales for the last year	$_____
Multiply by 0.2	$_____
Divide by 2	$_____
Divide by 5	$_____

The waste in an organization performing at approximately 3.5–4.0 sigma is
about 25 percent of gross revenue. This is where most North American companies are performing that do not have a strong program for improvement. When
an organization improves to 6.0 sigma level, the waste is less than 5 percent of
the gross revenues. Conservatively this represents 20 percent of gross revenue
available. Because some of the waste is going to be very difficult to eliminate,
we claim 50 percent of the identified waste, as a safety factor, so we divide the
total waste by 2.

It is unreasonable to think any organization can eliminate even half of the waste
instantly. It is going to take time to make the necessary changes, so we allow
a five-year effort to recover the 50 percent identified. On a straight-line basis,
that is 10 percent of the targeted waste each year.

The answer is simply 2 percent of gross sales added to the profit each year.
Sales times 0.2, divided by 2, divided by 5, equals 2 percent of sales as the

annual target, once you have Six Sigma projects delivering to the bottom line of the income statement.

Would you like to have that addition to your income statement each year? Do you currently have a way to deliver that kind of improvement? If the answers to these two questions are YES and NO, you are ready to implement a Six Sigma initiative! Now a number of people will demand to know what it is going to cost them to achieve these returns. The point is that these returns result AFTER absorbing all of the costs associated with Six Sigma methodologies. Profits are calculated after all expenses, including those that are part of the Six Sigma deployment processes, have been taken into account. It is important for organizations to understand the costs.

Before finalizing estimates, you should question how much you want each project to add to the bottom line of the income statement. We suggest that you consider as a minimum $100,000 on an annual basis. Some organizations require at least $250,000 and experience actual contributions per project of close to $1,000,000. Use the following equation when working with the $100,000 figure.

To estimate the number of Black Belts and Green Belts:

    2% of sales divided by $100,000
       (Number of projects per year)   _____
    Divide by 3 to get the number of Black Belts   _____
    Multiply by 10 to get the number of Green Belts   _____

Using the 2 percent of sales as the annual return, with each project returning $100,000, simple division obtains the number of projects you will need to complete each year. A Black Belt should complete between 2 and 4 projects each year. If you take the number of projects and divide by 3, you discover the number of Black Belts required to lead that number of projects. As an estimate, each Black Belt project will require 10 support team members. The Green Belt estimate for project support then is 10 times the number of projects. This assumes that all project team members are trained to at least Green Belt competence, and that there are no people serving on multiple teams. Now you have the number of Black Belts and the number of Green Belts required to achieve the improvements needed to return an additional 2 percent of sales to the bottom line of the income statement. If your organization has full-time Champions within a business, they must also be added to the cost. On a full-time basis there is customarily only one Champion per business. Smaller businesses may include Champions as a fraction of the senior-level manager's responsibilities.

Obtaining coaching, either external to the organization or internal (using Master Black Belts who have developed the necessary expertise, ability, and competency), is a matter of preference. Either choice results in one full-time coach for every ten to twenty-five Black Belts. Coaching for the management and Champions as a rule comes from outside the organization.

Many organizations that have successfully implemented Six Sigma require all managers and future managers to be trained to at least a Green Belt level of competence. This is not an immediate obligation. We recommend considering these training costs as part of the ongoing employee development cost, tracking these training costs, and adding them to the total cost of Six Sigma deployment. Compared to the gains realized from the successful deployment of Six Sigma, these costs will be small, no matter which approach is used.

The major costs associated with Six Sigma are:

- Salaries for the Black Belts (Black Belts generally command 10–40% premium over their base)
- Salaries for the Green Belts (the fraction of time they spend working on the project is usually 10–20%)
- Overtime or other arrangement for the work Green Belts normally do while they are doing project-related work
- Training cost for Black Belts ($15,000–$50,000 including travel, less if done in-house)
- Training cost for Green Belts ($5,000–$10,000 including travel, less if done on-site)
- Management and Champion training cost ($12,000–$15,000)
- Awareness training for the organization ($200 per employee)
- Coaching for Black Belts and management Champions ($1,000–$2,000/day)
- Support for projects:
  - Computers (usually a top-end laptop with access to printers)
  - Software (the full Office package plus a good statistical package)
  - Travel (project-dependent)
  - Office staff (project-dependent)
  - Office space (standard for valued professionals)

From the above list, you can make a budget estimate of the costs associated with deploying Six Sigma in your organization.

No matter what the costs turn out to be, the real questions are: Would you like to have an additional 2 percent of gross sales added to net income? Do you have another way that you reasonably expect to make that kind of improvement to income?

At this stage, most organizations are ready to implement Six Sigma methodologies. The question then becomes, how do we deploy Six Sigma to ensure that we get the kind of bottom-line gains calculated?

Some Advice Regarding Training

The training costs for Black Belts range from very high to very low, and the time needed for training ranges from one to four weeks. We encourage you to look at the content of the training. If training is exclusively statistical in nature, you will miss some very important skills for managing human relationships. In fact, one of the common failings we have seen with Black Belts training programs is a rigorous focus on technical competence and on the statistical tools and techniques. Black Belts are often not accustomed to dealing with persons less educated than they are, and they generally find it seriously challenging to lead teams made up of less-educated subject matter experts. On the other hand, statistics are very important, and trying to skim through the training in just a few days does not allow for adequate understanding or the ability to apply the statistical techniques. There is considerable evidence that a four-week training program for Black Belts works very well. It usually comprises one week of dedicated training followed by three or four weeks of applying the material to a real project. It is essential that the Black Belts do not return to their previous work responsibilities, but rather focus all their energies on completing the training. Management should forge the Black Belt position into a full-time job. Having Black Belts work only on projects leaves time during the training process for application and review of all required materials. If a particular topic becomes difficult to master, the Black Belt will also have the time to dedicate to the resolution and understanding of that skill requirement. Applying this approach allows your Black Belt to conclude the training and be able to perform the tasks required to lead a successful Six Sigma project. Do not be fooled by those who promise a one- or two-week training blitz. Unless candidates are extremely well prepared, knowing most of the material in advance, these classes will provide little more than an understanding that the tools and techniques of Six Sigma can be applied successfully. They will remain incompetent at the performance of the skills required to attain the organization's desired strategic business results. The price range for the four weeks of Black Belt training varies widely. The highest price or lowest price is not necessarily the highest value. Our advice is to check the curriculum content for a balance of both human relations/team skills and statistical skills. Much the same cautions are appropriate for Green Belt training curriculum and processes.

5

MANAGEMENT AND CHAMPIONS

Management gets what management wants. Some may disagree with this statement, but we contend that when management does not get what they want, changes are implemented to produce actions toward progression in the direction they desired. Of management's many hopes and wishes, for some they seem unwilling to pay the price in terms of change, effort, personal involvement, time, money, people, or other resources. These hopes and wishes are likely to fall into the category of "nice to have as far as the management is concerned." Lip service is provided, but real involvement and commitment are missing. Unfortunately, some in management will allow the destruction of an organization or company before forging the changes needed to sustain a viable and vibrant enterprise. In these cases, we contend that management wants to avoid change more than they want to ensure the viability of the organization. Some managers may not know what kinds of changes are required. Nevertheless, if management is serious about ensuring the viability of the organization, Six Sigma deployment will provide data that specify the particulars of the required changes.

There is almost unanimous agreement that for any effort to succeed the single most important ingredient is management leadership. Leadership is much more than just support, delegation, or the assigning of resources. Personal direct involvement and personal time are probably two of the more consequential measures of any leadership initiative.

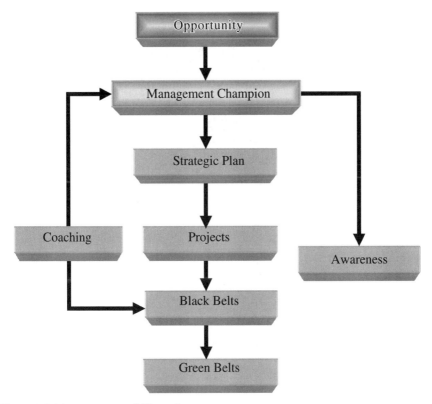

Figure 5.1 Management and Champions.

In our deployment flow chart we are at the Management Champion box
(Figure 5.1). Within the Six Sigma Management/Leadership area, a number of
labels are assigned to the various roles and responsibilities of these functions.
In current Six Sigma literature, there seems to be little consistency among
those various roles and responsibilities. We will share with you our definitions of some common titles, roles, responsibilities, and duties for Six Sigma
Management and Champion leadership. Not every organization may have all
of these titles; smaller organizations often combine them into one job function.
Management's and champions' roles, responsibilities, and duties are essential
for deployment of Six Sigma. Please do not be too concerned about the titles.
The roles, responsibilities, and duties are of the most importance.

EXECUTIVE QUALITY COUNCIL

The Executive Quality Council includes the people who predetermine strategy for the organization. Typically, the council is made up of six to nine

Chief (X) Officers of the organization; (X) equals job functions including executive, operating, financial, research, legal, and so on. In short, these people run the organization. Added to the Quality Council may be support executives holding responsibility for implementing the Six Sigma initiatives across an organization. Leadership is accountable for the success of Six Sigma implementation. Leadership cannot make the mistake of delegating the responsibility to others. It is the responsibility of senior management to make the changes, in the organizational behavior and culture, necessary in Six Sigma deployment. Using the DMAIC process and metrics associated with Six Sigma, one might legitimately expect to experience certain short-term improvements. This will not, however, transform an organization into a world-class performer! Only those that are world-class performers can prosper in the current global marketplace, given the competition advancing from unanticipated areas.

Changing the way we think is a first step towards executing world-class performance. Most of us find producing change difficult and time-consuming. Additionally, as managers, we often fail to understand that these same changes must occur with every other individual within the organization, and that they too need time and effort to understand the changes. Management can foster this process by exhibiting their commitment and by using a common language.

We suggest that organizations demonstrate commitment by taking the following steps:

- Document and manage processes end to end. It is important to assign responsibility to guarantee cross-functional management of critical processes.
- Clearly define and regularly update your Customer's Critical Criteria. The measurements of your process outputs, process activities, and process inputs must be thorough and meaningful.
- The organization managers, process owners, and other stakeholders should use the measurements and process knowledge to assess performance in terms of real time and to take corrective actions that address problems and opportunities alike.
- Process improvements should be built around the improvement tools of Six Sigma and used to constantly improve the organization's performance levels, competitiveness, and profitability.

Once the organization's strategic plan is refined, the Executive Quality Council is in a position to complete the following steps:

1. In Step One, develop clear comprehension of how your organizations "fits" together by using a high-level process map. The senior

management team creates a map or an inventory list of "value-delivering activities" within the organization. The senior management team questions:

- What are our core or value-adding processes?
- What products and/or services do we provide to customers?
- Which key customers do they serve?
- Interfaces and handoffs between groups are explicit and well managed. How do the processes "flow" across our organization?
- What are our strengths?

2. In Step Two, senior management develops crystal-clear and complete descriptions of the factors that drive customers' satisfaction for each output and process requirement or specification. The senior management team questions:
   - What do we really comprehend about our customers?
   - Do we focus on service and output requirements?
   - Do we have output requirements tied to the end products/services that make them work for the customer?
   - Do we have service requirements describing how our organization interacts with our customers?
   - Do we understand our competitor's customers?
   - Do we have effective, broad-based input from the customers and market inputs?
   - What are our Customer's Critical Criteria? What are the requirements and enhancements?
   - Have we translated customer responses into crystal-clear requirements or specifications?

3. In Step Three, senior management develops a system of current performance measurement. The senior management team questions:
   - Do we have established baseline measures that quantify evaluations of current process performances?
   - Are we accurately measuring our performance against customer requirements?
   - Do measures encompass both service and output specifications?
   - Have we established capability measurements that assess the ability of the current process or output to deliver Customer's Critical Criteria?
   - Do these measurements include sigma scores for each process that allow for comparison of very different processes?
   - Have we developed new or enhanced measurement methods and resources for ongoing measurements against customer-focused performance standards?

- Are these data easily accessible?
- How well do we use the data to evaluate and fine-tune our process performance?
- Do the people working in our process comprehend the measures?
- Do they grasp what to do with the measurement data?
- Do we have input or process measures in place to help us observe potential problems or identify opportunities before they happen?

4. In Step Four, senior management develops an inventory and strategy of improvement priorities. The senior management team develops potential Six Sigma projects assessed according to their impact and feasibility. They develop process improvement solutions targeted to specific root causes of existing opportunities, or creates new activities or workflows to meet new demands or new technology, or to achieve dramatic improvements in speed, accuracy, cost, or performance. The senior management team questions:
- Do we have critical problems or opportunities needing attention?
- Can we identify the likely payoffs for resolving these problems or opportunities?
- Do we have the necessary resources deployed for tackling these problems?
- Are we attempting to solve problems with a Band-Aid approach?
- Is there a clear, proactive process in place to develop root cause analysis and focus solutions efforts?
- Are we willing and able to design or redesign processes when we have determined that our current design is no longer viable?
- Are our essential leaders engaged in and actively supporting improvement efforts?
- Are we measuring results and ensuring that solutions meet payoff criteria?

We recommend that the Executive Quality Council address the following questions before starting Six Sigma deployments: For what reason and on what level will we communicate with our organization about Six Sigma, its impact on our strategic objectives, progress, successes, and so on? What does process thinking mean to us as an organization and as individuals? How will we encourage innovation within the organization? When improvements are made, how will they be sustained? (Who is responsible for sustaining improvement? What are our checks? What is our metrics for improvement? What is our reward system?)

Using the language of Six Sigma in a logical and meaningful fashion will encourage the organizational team members and demonstrate the seriousness of your process undertaking. Members of this council must take time to address the start of each Black Belt training cycle and conclude final class day, thus demonstrating executive management commitment to Six Sigma process, methodologies, and language in a meaningful way.

Every member of management must be comfortable in communicating the following items to the rest of the organization:

- Mission, vision, and vision elements of the strategic plan
- World-class performance
- Customer's Critical Criteria
- Process management
- Conversion of data into information
- Relation of sigma levels to defects per unit and defects per million opportunities
- Rolled-through yield
- Statistical process control
- Concepts of design of experiments, including interactions
- Cycle time
- Pareto Principle

We have included a glossary of terms used in this book, and others, to support your efforts in this area. You will find multiple definitions for some of the terms, which demonstrates the diversity of current thinking. There is no substitute for actual experience. For example, have you ever taken a table of numbers and constructed a histogram and/or a control chart? This personal experience probably revealed to you the power of converting data into information and provided you with the ability to speak with authority and conviction on the topic.

PROCESS OWNERS

Process owners are responsible for a specific process from the beginning to the end. Major processes will cross multiple functional and departmental boundaries. Some organizations have established process owners well before they start to consider Six Sigma deployment, and others do not have this responsibility in place. Subsequently, Six Sigma projects focus on process improvement and sustaining gains, requiring this role to be filled. One of the meaningful concepts for process owners and management to grasp is that the process is no respecter of boundaries. The process does not care if boundaries are internal to your company or if they are external and between companies. Some processes may enter and leave your organization or company several times. It is not unusual for the customer to be a supplier. As we have said earlier, you have to get to the details, and "The devil is in the details."

Effective process owners will often become involved in the supply chain to their process. The voice of the customer must travel to all of the appropriate places in the process, so that everyone involved will discern the Customer's Critical Criteria. In this manner, boundaries within the company will become more process-friendly and less inaccessible. Within each major process, there are likely to be any number of subprocesses. It is conceivable that a specific level of management will own the high-level processes. Identifying and comprehending subprocesses is likely to uncover owners who may not be members of management or supervision. Astonished? An empowered hourly employee can certainly be the owner of a subprocess. Process owners are always accountable to the business.

We suggest that process owners' responsibilities include:

- Comprehending the entire process from end to end, which entails (1) maintaining a current process map from suppliers through customers, and (2) understanding the Customer's Critical Criteria for the process, including the required and enhanced criteria
- Maintaining and monitoring process performance metrics
- Benchmarking the process
- Sponsoring Black Belts for process improvement projects
- Participating in, or providing other proficient people for, projects that affect the process
- Comprehending the linkage of this process to other processes
- Sustaining the gains achieved through Six Sigma projects
- Sponsoring and participating in continuous improvement efforts for the process
- Working with suppliers and customers of the process for mutual improvement
- Providing suggestions regarding proposed changes brought about by Six Sigma projects

CHAMPIONS

The Champions report directly to the business leadership team. Certain Champions may serve on the Executive Quality Council. Champions are responsible for the overall success of Six Sigma processes.

We suggest that Six Sigma Champion duties include the following: Champions represent Black Belts directly to the organization's senior management team. Access to management is often one of the most prized recognitions. Champions furnish guidance and counsel for Black Belts, most of whom are junior managers in the organization and enjoy considerable opportunity ahead of them. Champions actively assist Black Belts in the course of leading the organization's value-added projects. Champions also obtain the necessary resources for Black Belts, including technical resources. They maintain an adequate budget to cover travel, computers, software, training, development, office space, phones, fax, e-mail, Internet, and so forth, for Black Belt project managers.

Even as projects are budgeted, some needs remain unanticipated. Champions are obligated to facilitate Black Belts in performing their jobs successfully. Champions frequently confer with senior management as to career path and advancement opportunities for Black Belts, and assist in identifying potential Black Belts. They are also responsible for identifying the next two potential replacements for each Black Belt.

Champions are two or three projects ahead of each Black Belt, to ensure that these people enjoy value-added productive work. One method to accomplish this is to anticipate the next project assignment for Black Belts. Champions substantiate that Black Belts have an opportunity to apply a diversity of skills by matching projects and tools with each, in order to expand individual growth potentials. Most people use and reuse the approaches that have been successful for them in the past. There is nothing wrong with building on success, but very few projects will afford the convenient opportunity to use all of the Six Sigma tools and techniques available. If we do not use, or we ignore, techniques and skills, we will lose them.

Champions maintain a reserve of ensuing projects for Black Belts. Projects may be formal or informal in nature. Champions accumulate the ideas and do a quick assessment of the potential value expected. Performing a perpetual project review exposes changing conditions. Reviews may provide discovery that a desirable project is not viable or that what was felt to be an unattractive project may in fact be a profitable one. One calamity, in our view, is to have consumed resources to instruct Black Belts but not have potential projects identified for assignment. We hear rumors of organizations that have experienced this dilemma, which demonstrates that the basic preparation by management was incomplete before they rushed into training. Champions are aware of and prepared for project cancellations due to changes in the needs of the business, or for projects that take less time than anticipated, or that no longer meet the current Customer's Critical Criteria.

Champions review Black Belts' project management activities and progress. There are logical points at which to review each Six Sigma project, such as each step in the DMAIC process. Project reviews help keep the Champion in harmony with the project's status. Black Belts may depart from a project for any number of reasons, such as quitting the company, a floundering project, death, and so on. The Champion who is abreast of the project status can greatly assist the new Black Belt in the project's forward movement and success. It is common for projects to cross paths within the organization. This often provides Champions with the opportunity to coordinate projects with other stakeholders. Project crossings have the potential of creating conflict between the project team members, sponsors, and/or managers of the processes, functions, or departments. Communications, coordination, and planning can often mitigate these issues.

Champions sponsor presentations of completed Black Belt projects to senior management. Presentations to senior management provide recognition of Black Belts. These presentations also furnish senior management with an active opportunity to connect with the people who, in our opinion, are the pool from which future leaders of the organization will emerge.

Champions confirm by means of accounting, or other project evaluation methods, that the gains claimed by a completed project are real and have added profit to the bottom line. They halt projects that do not have the potential for the gains required of a Six Sigma project. Some projects may only be appropriate for continuous improvement efforts by a specific function or department, as they may not meet the criteria for a Six Sigma project.

Champions collect ideas for future projects from the organizational stakeholders. They sponsor and share learning experiences between all the organization's Black Belts. Seeing and hearing about peers' success stories and/or less than anticipated benefits serves to enrich the collective spirit and skill set of each team member. Inspiration and growth opportunity take place in both success and setback stories when shared by peers. Champions also augment individual Black Belts' skill sets by exposing them to seminars, books, videos, speakers, and the like regarding Six Sigma methodologies.

Champions have essential responsibility for assessing the skill level and for challenging each Black Belt's Six Sigma methodologies proficiency level. Champions must also consider the needs of the organization when making these assignments. An additional challenge for management is to balance the skill level and assignments for all employees. Properly balanced assignments create job interest, pride, ownership, commitment, enjoyment, and motivation for those persons involved. When the skill level far exceeds the challenge of the assignment, boredom is often the result. When the challenge of the assignment is far above the skill level of the employee, anxiety is often the result.

In a seller's market (as it is for Black Belts), when boredom or anxiety is the result of assignments, employees often will seek work elsewhere.

Sponsors

Sponsors are the persons who propose a project for selection and assignment. Sponsors generally have the budget authority to pay the expenses affiliated with the Six Sigma project team's efforts. We suggest that sponsors customarily provide the following:

- Problem/concern statement or statement of the opportunity
- Contact with process owner(s)
- Functional or department contacts likely to be affected by the project
- Team members or subject matter experts for the project
- Project constraints or boundaries
- Suppliers to the process
- First step in the process
- Customer's Critical Criteria of the process
- Last step of the process
- Intermediate steps of the process
- Constraints, boundaries, and limits, whether in scope, time, or other resources

Sponsors frequently provide very large and subscoped projects. Black Belts and Champions spend appreciable time with prospective Sponsors in defining a properly scoped project. It is not extraordinary for two or more projects to extend from a Sponsor's original concept. We will discuss project selections and assignments later. Remember, a firmly defined project will have a much higher probability of succeeding than one that is subscoped.

Master Black Belts

Customarily, Master Black Belts have abundant experience as successful project managers. They possess a depth and breadth of Six Sigma philosophy, methodologies, and skill sets greater than those of Black Belts. Master Black Belts tend to remain in their roles for longer periods than do Black Belts. Larger projects commonly require a Master Black Belt for overall leadership, supported by a group of Black Belts who lead subprojects within the overall project scheme.

It is highly desirable for a Master Black Belt to be one who can lead in strategic development, group facilitation, fundamental statistical tools, project management, behavior modification, SPC, design of experiments, data mining, management tools, presentation skills, and procedure writing. The majority of Master Black Belts concentrate on a specific area and possess strong competence in various others. We have discovered there are two general qualities that characterize Master Black Belts:

- Master Black Belts are more expert in the soft people skills.
- Master Black Belts have a strong statistical expertise.

We suggest that Master Black Belts have the ability to perform equally well in both areas. We have discovered, however, that some Master Black Belts may need additional assistance in specific areas.

We recommend that organizations consider maintaining one Master Black Belt to ten Black Belt project managers. Some organizations insist that the Master Black Belts be extremely proficient in the statistical analysis area, and they tend to lean on Champions to provide the soft skill expertise, as need arises. We recommend that organizations evolve, groom, and promote Master Black Belts from within. Black Belts typically spend two or three years on assignment and then return to the line organization. The Master Black Belt function creates a "technical ladder" career step for those who become Black Belts and finds that this is their life calling. Organizations cannot allow the ratio of Master Black Belts to Black Belts to get out of hand. These valuable employees continue to enrich their personal knowledge, capabilities, and advancement of Six Sigma deployments, as well as add value back to the organization.

We recommend that Master Black Belts' responsibilities include:

- Serving as project managers for very large Six Sigma projects
- Facilitating the continuing education of the organization's Black Belts and Green Belts
- Coaching Black Belts on their projects (see Chapter 8, "Coaching")
- Hosting and facilitating Black Belt sharing sessions
- Advancing the organization's capability within the boundaries of Six Sigma philosophies, methodologies, techniques, and tools
- Participating in the career planning of the Black Belts they coach

Not every organization needs an in-house Master Black Belt when it is first deploying Six Sigma. An outside consultant can perform many of these duties, or some of the expertise can be picked up from a supplier or customer who is more advanced with Six Sigma.

Black Belts

We will spend time on Black Belt project managers' training later. Ordinarily, Black Belts will possess a technical background, which helps them to understand and use the appropriate statistical techniques. The Black Belt function is a full-time job lasting about two or three years. The following section outlines the issues that management should take into consideration before naming the first Black Belt.

We recommend that all Black Belts complete an organizationally approved training program of Six Sigma process philosophies, methodologies, and tools. A few Black Belts may already have competence in some of these areas.

Black Belts need, as a minimum, an all-purpose PC-based statistical software package. We recommend that organizations select a common PC-based statistical software package for the entire organization. Ensure that the PC-based statistical software package purchased is capable of doing basic Design of Experiments and Statistical Process Control (control charts). There are several good single-purpose PC-based statistical software packages. Most vendors will grant discounts based on relatively small volume needs.

We suggest that Black Belt candidate criteria should allow only high-performance people who have a demonstrated capability to produce. Our personal preference is that those considered for future promotion to management-level positions serve first in the capacity of Black Belt project managers. Unexpected dividends often result, and it sends a clear message through the organization that Six Sigma is important. This is a pathway into management for those who aspire to higher levels. We recommend that management consider people for Black Belt duties who are comfortable dealing with advanced mathematics. The computer and software programs we spoke of earlier will do most of the grunt work, but statistical concepts will be difficult to understand and apply without a good mathematics foundation.

Black Belt assignments are generally for the duration of two to three years. Project completion during this time will recover all training and support costs, and provide a commensurate return on the investment. Project work can be very demanding, and there is a potential for burnout. We suggest that management consider a two-year minimum and five-year maximum as acceptable ranges for Black Belt service.

We feel that organizations must reward Black Belts quite well and without apology. Black Belts' projects will return multiples of their compensation to the bottom line of the organization. We recommend that Black Belt career planning include considerable investment in continued training, personal career coaching, and promotion back into the line organization. Suggested compensation plans for Black Belts include a fraction of the project savings, base pay plus a specific percentage increase for completing the Black Belt training, pay increases when two projects are successfully completed, and executive stock plans with varying levels of stock options.

GREEN BELTS

Green Belt employees acquire, through training and experience, a level of competency using the Six Sigma philosophy, methodologies, and tools. We recommend that organizations train and qualify both management and project team members as Green Belts. In our opinion, every manager needs to become Green Belt qualified. We suggest that management consider the following reasons:

- Managers may be Sponsors of projects.
- Managers may be process owners.
- Managers may be on project teams.
- They will manage the changed organization.
- Managers may be department or function leaders affected by Six Sigma projects.
- Management must support the process thinking approach.
- Managers must gain general knowledge and skills regarding the use of the Six Sigma techniques, tools, and methods of data reporting.
- Managers must gain general knowledge and skills regarding development and motivation of project team members.
- Managers use data-based decision making and reporting.
- Management supports the changes brought about by Six Sigma projects.
- Managers continuously improve their areas of responsibility, using the tools and techniques on efforts smaller than full Six Sigma projects.

We do not recommend Six Sigma deployment programs requiring all managers to attend Green Belt training early on. We do suggest the following step-by-step approach, which has worked well for other organizations:

- The organization produces a formal notice that offers Green Belt training for those wishing to participate. An unannounced paid cash bonus is provided for those who complete the training within the first one to five offerings.
- Next, the organization produces a formal notice regarding payment of cash bonuses to all managers who complete Six Sigma training by a

certain date. This bonus is smaller than the one suggested in the first step.

- Third, the organization sends a formal notice to all managers remaining unqualified as Green Belts, informing them they will not be eligible for promotion beyond their current position until they complete Green Belt training.
- Finally, the organization produces a formal notice announcing that any employee who remains unqualified as a Green Belt will not be eligible for promotion beyond his or her current position.

Some steps may not be possible for smaller organizations. We do strongly recommend that managers attend Six Sigma training that is exclusively for the organization, as opposed to training for the public at large, because, in this way, there is no compromising of confidential or proprietary information among participants, and stakeholders are free to discuss specific organizational issues and concerns. This approach also allows for Six Sigma training customized to the specific organization and industry. It enables managers to work together on projects and to maintain vigilant focus on the organization's strategic plan and objectives.

PROJECT TEAM MEMBERS

Subject matter experts serve on the Six Sigma projects led by the Black Belts. Effective project team members must be able to understand and use Six Sigma tools, techniques, and approaches. Perhaps a more fundamental rationale for team members to experience Green Belt training is to gain an understanding of "what is in it for me." It is human nature for each of us to ask why we are included in some activities. Leaders must recognize this basic human need, address it adequately, right up front, then move on to effective team building. We recommend that leadership assign project team members to specific projects before providing Green Belt training. We also recommend that the Black Belt project manager attend all or most of the Green Belt training presented to the project team. This training experience allows the project manager the opportunity to learn about the team members' behaviors, skill abilities, and motivations. The training setting provides many valuable interpersonal relationship interactions between all team members and the project manager.

We suggest the following Green Belt project team membership issues to consider:

- Who is responsible for selecting each potential Green Belt for a specific project?

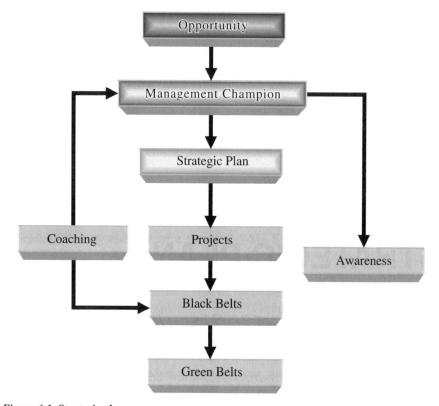

Figure 6.1 Strategic plan.

context or fabric to understand how the organization has developed to this point.

V. Critical Issues. Included here is a discussion of the most significant issues facing the organization. This includes internal as well as external factors.

VI. Vision Elements. Usually limited to six to nine, these are the major areas where the organization will focus its attention. They are very high-level, and if performance in each of these areas is exceptional, the organization will be well on its way to reaching the vision.

VII. Key Performance Indicators. These are important, relatively high-level measures that track performance in each of the vision element areas. They are of greatest value when historical performance, current performance, and desired levels are all shown together.

VIII. Objectives. Within each vision element there may be several objectives. These are the things that the organization would like to achieve within the specific vision element.

IX. Goals. Each objective will have set goals. There may be several goals for each objective. Each goal should be SMART (specific, measurable, action-oriented, realistic, and time-bound).

X. Plans. These are the specific details, down to the task level, that will ensure, when accomplished, that the goals are reached and the objectives met.

XI. Appendices. This is the place for important information that is not included elsewhere, but should not be left out. Some choose brevity and have few documents in this section, while other strategic plans present an extensive amount of information in the appendices.

The strategic plan process entails:

- Examining the organization's critical issues
- Determining how the organization's strengths and skills can be employed to address the critical issues
- Analyzing opportunities and strengths and looking for ways to synthesize the two
- Exploring and choosing the best approaches for the organization

Finally, strategic planning, though described as disciplined, does not typically flow smoothly from one step to the next. It is a creative process, and the fresh insight arrived at today might very well alter the decision made yesterday. Inevitably, the process moves forward and back several times before arriving at the final set of decisions. Therefore, no one should be surprised if the process feels less like a comfortable trip on a commuter train than like a ride on a roller coaster. However, even roller coaster cars arrive at their destination, as long as they stay on track!

A strategic plan is (1) the organizational actions in (2) the marketplace that produce (3) a competitive advantage. If all three elements are not present, then we contend that a working strategic plan is not in place. There are many ambitious statements that may direct the development of a strategic plan, but until you can define the organizational actions you intend to take, the marketplace where those actions will occur, and the competitive advantage that will result from taking those actions in the defined market, you do not have a strategic plan.

Strategic planning is not a mysterious effort that only the select few can understand. In essence, it is quite simple:

- What are our values, beliefs, and purpose?
- Where are we today?
- What is success?

- What is the gap?
- How do we address the gap?

A strategic plan has multiple uses. Primary among these is to ensure that the leadership and the organization as a whole have a common understanding of the direction and importance of what is happening within the organization. It serves as an alignment tool, focusing attention, direction, and effort toward some common results. The strategic plan enhances communication efforts and gives guidance as important decisions are being considered. The strategic plan is one of the few duties that senior leadership cannot delegate to anyone else. Staff, consultants, and others may assist and provide some of the detail work, but the senior leadership must own the strategic plan. While the final documents are important, it is the process of developing the strategic plan that is most important. The discussion of views and values by the leadership and feedback from the organization all help to weave the fabric of the strategic plan, which defines what the organization is really about.

The very reason for an organization to exist is captured as part of its strategic plan. Typically, this is considered the mission of an organization. Its long-term objectives are stated in its vision. Frequently, visions are 10, 20, 50 years or even longer in the future. Unless there is agreement on these two critical perspectives of the organization, it is difficult for managers and employees to know how to act and behave. Tom Peters identified a clear vision of the desired future state of the organization as an essential component of high performance. The strategic plan not only gives direction but also defines what is outside of consideration.

Once the mission and vision are defined, a number of key areas are identified. These are the fundamental objectives of the organization, or key performance areas. Each key performance area will have one or more key performance indicators that allow for tracking of progress. Within each objective area will exist a number of goals, and plans to achieve those goals. Each goal will have measurement criteria that are specific to it and support the key performance indicator. The decisive test for the objectives is the question: If all of the objectives are met and the desired level for each key performance indicator is achieved, will the organization move toward its vision and be actively accomplishing its mission? If the answer to this question is anything other than an unqualified YES, then it is time to return to work on the objectives or redefine the vision and mission. Unless there is an unqualified yes, there is no consistency among the vision, mission, and objectives.

Six Sigma is not the vision or mission of the organization. Achieving a level of performance where the numbers of defects are six (6) sigma or less may be one of the key performance indicators. The process of implementing Six Sigma may be one of the plans, but Six Sigma is not, even in the broadest definition, sufficient to be the strategic plan for an organization. Six Sigma CAN make a major contribution to achieving the strategic objectives of the organization in a time frame not achievable in any other way.

DEFINING THE VISION

Our conviction is that the vision of an organization is not dreamed up from the mist, but instead has it roots in the values, beliefs, principles, and philosophy of the leadership. In preparing of the vision, there are a number of questions that the leadership team must ask themselves, and they must come to total agreement about the answers. These questions include:

- Why does this organization exist?
- What are the long-term objectives of the owners/directors?
- What are our values? (What is important? What is not important?)
- What are our principles?
- How do we relate to other stakeholders (suppliers, customers, employees, owners, community, government, competitors)?

There is nothing wrong with using some of the thoughts and ideas of great thinkers and organizations over the ages to help in these exercises. Religion and philosophy are totally appropriate places for inspiration. Our only caution is that the entire team must be willing to have the organization live the principles and values espoused. Better to leave high-sounding words and ideals out than to have them included and not followed.

One of our fundamental beliefs is that we should be honest and demonstrate integrity. These frequently get confused and unfortunately seem to be disappearing in some organization as they chase after the almighty dollar. Individuals can be honest and have no integrity, or demonstrate integrity and not be honest. These are unusual conditions, and people and organizations that have difficulty being honest usually have little integrity, and vice versa.

> Honesty. The actions come first, and then the words. The words are consistent with the actions.
> "I did chop down the cherry tree."—George Washington
>
> Integrity. The words come first, and then the actions. The actions are consistent with the words.
> "I will not run for a third term as President."—George Washington.

For all too many, it is difficult to face the consequences of their actions or live up to the commitment of their words. It may be worth a little of your time to reflect on your personal performance and that of your organization. Is there unambiguous evidence of honesty and integrity? Are there specific instances that might cause some to question the honesty or integrity of you or your organization? The past cannot be changed. What can change are the words and actions from this point forward. Without honesty and integrity we do not know how a meaningful strategic plan can be developed or implemented; the fundamental foundation is missing, and a house built without a foundation will collapse.

Once there is agreement on these issues, a vision statement can be written. The vision is an image of the purpose of the organization. As a recommendation: once agreement has been reached on the above questions, have a participant take all of the input and notes from the discussion and write a draft vision statement. Once this statement is written, the leadership will need to review and modify it so that all can agree. Writing as a group effort is usually not a very productive or efficient exercise. Much better to follow the model established by the framers of the Declaration of Independence and have a member create the draft, then have the entire team make appropriate changes.

A well-written vision with a sharp focus and clear message will serve better than a rambling statement that includes all the superlative adjectives known. The vision should of course have considerable stretch and generally exceed our grasp, but if it is so general and vague that it provides no direction or guidance in decision making, it provides no useful function. Develop something that people in your organization can look to in times of extreme stress and uncertainty to give them guidance in making totally unanticipated decisions.

EXTERNAL AND INTERNAL ASSESSMENT

This activity includes conducting a scan or review of the organization's political, social, economic, and technical environment. Planners carefully consider various driving forces in the environment, such as increasing competition and changing demographics, for example. There are a number of ways in which to do assessments. One of the most popular is to do Strengths, Weakness, Opportunities, and Threats (SWOT) analysis. The idea of this type of analysis is that an organization will:

- Build on strengths
- Resolve weaknesses
- Exploit opportunities
- Avoid threats

SWOT Analysis is best done in a structured fashion, considering specific areas. Data should be sought to support any opinions or judgments expressed.

This is especially true in the areas that are considered strengths. "In God we trust, all others bring data."

Strengths are positive aspects internal to the organization:

- What are your advantages?
- What do you do well?

Consider this from your own point of view and from the point of view of others. Don't be modest, don't indulge in fantasy, be realistic. If you are having any difficulty with this, try writing down a list of your characteristics. Some of these will hopefully be strengths!

Could a competitor emulate your competency? Can customers move elsewhere seeking satisfaction?

Is this strength being neglected, and could it deteriorate over time? Did it take time and resources to develop this strength? Is the investment in developing this strength largely irreversible?

The things we do best of all are:

- _____
- _____
- _____
- _____

Weaknesses are negative aspects internal to the organization:

- What could be improved?
- What is done badly?
- What should be avoided?

Again, this should be considered from an internal and external basis—do other people perceive weaknesses that you do not see? Do your competitors do anything better than you? It is best to be realistic now, and face any unpleasant truths as soon as possible. Will this weakness improve when the underlying causes are tackled? Is this a weakness because the organization lacks a basic requirement? Is this weakness striking at the core of the organization's business?

The things we are worst at are:

- _____
- _____
- _____

Opportunities are positive aspects external to the organization. Where are the good chances facing you? What are the interesting trends? Useful opportunities can come from such things as:

- Changes in technology and markets, on both a broad and narrow scale
- Changes in government policy related to your field
- Changes in social patterns, population profiles, lifestyles, and the like
- Changes in the market
- Local events

Is this an area that the organization could enter almost immediately, by reconfiguring itself, to offer value to a market? Is this a potential area the organization could enter in the future, provided it developed towards this objective?

If we could take advantage of the following, it would propel us ahead of everyone else:

- _____
- _____
- _____
- _____

Threats are negative aspects external to the organization. What obstacles do you face? What is your competition doing? Are the required specifications for your job, products, or services changing? Is changing technology threatening your position? Do you have bad debt or cash flow problems?

What are our threats?

- Short-term
- Medium-term
- Long-term

If someone were able to do the following, it would put us out of business:

- _____
- _____
- _____
- _____

SWOT analysis can be done on a wide number of different areas and is not limited to strategic planning. In fact when Six Sigma project selection is started, we recommend SWOT analysis for every project before it is formally assigned to a Black Belt. Some areas where SWOT may prove fruitful include:

- Advertising
- Capacity
- Community relations
- Competitors
- Convenience
- Coordination
- Costs
- Customers
- Energy
- Finance
- General economy
- Image or reputation
- Inventory
- Labor
- Location
- Management Information Systems
- Market share
- New markets
- New products
- Operations
- Organizational structure
- Packaging
- Planning
- Product line
- Productivity
- Public perception
- Quality
- Raw materials
- Regulations
- Research and development
- Safety
- Sales
- Security
- Service
- Staffing
- Succession
- Supervision
- Supply chain
- Technology

- Training
- Versatility
- Work force capability
- Other areas specific to your needs

For each area, a simple form can be used to collect the SWOT analysis, as shown in Figure 6.2. When the SWOT analysis is completed for the areas deemed appropriate, there will be a range within each of the quadrants. Many

Area/Project/Topic: _____

Strength—Advantage of opportunities or reduced impact of barriers

Weakness—Factors in the way of opportunities or reduced impact of barriers

Opportunities—Outside factors that allow or encourage action

Threats—Factors in the way of opportunities and encouragement of success

Factors	Internal	External
Positive	Strengths	Opportunities
Negative	Weakness	Threats

Figure 6.2 SWOT data collection form.

find it useful to divide the strengths, weaknesses, opportunities, and threats into four separate documents. By doing this, it is possible to assess the relative strengths, weaknesses, threats, and opportunities. In order to accomplish this, a rating scale is required. This is also a good place to document some of the reasoning for each item. In Figures 6.3, 6.4, 6.5, and 6.6 we offer tables that

Scale:

1 = Better than the average competitor

2 = Top 25–50%

3 = Top 25%

4 = Top 10%

5 = World-class

Item	1	2	3	4	5	Why

Figure 6.3 Strengths.

Scale:

1 = Slightly worse than the average competitor

2 = Bottom 25–50%

3 = Bottom 25%

4 = Bottom 10%

5 = We stink.

Item	1	2	3	4	5	Why

Figure 6.4 Weaknesses.

Scale:

1 = Incremental improvement in one area

2 = Incremental improvement in multiple areas/departments/functions

3 = Significant improvements in one area only

4 = Significant improvements in multiple areas/departments/functions

5 = Change the future of the organization

Item	1	2	3	4	5	Why

Figure 6.5 Opportunities.

can be used for this effort. It is easy to then construct a Pareto Chart for each of the four areas, using the consistent rating scale. Strengths are shown in Figure 6.3, Weaknesses in Figure 6.4, Opportunities in Figure 6.5, and Threats in Figure 6.6.

With the internal and external analysis completed, the leadership should be in a position to draft a mission statement for the organization.

Scale:

1 = Slight negative impact

2 = Negative impact on multiple areas/departments/functions

3 = Significant challenge to the organization

4 = Major damage to the organization as a whole

5 = Destroys the organization

Item	1	2	3	4	5	Why

Figure 6.6 Threats.

DEVELOPING THE MISSION

The mission is what the organization does. If the mission is executed to per-fection, then the vision should be achieved. A mission can be written for every level in an organization and for each individual within that organization. High levels in the organization will have a more general mission. At the individual level, the mission can be very task-specific. Here we are going to focus on the mission for the organization as a whole. The mission statement clearly states why the organization exists: it exists to achieve some purpose or objective. Given the values, principles, and vision of the organization, the mission is how the vision will be accomplished. As with the vision statement, we strongly recommend that someone be assigned to prepare a draft mission statement and then return to the leadership team for editing. Clarity and understanding should be the goal—not the inclusion of every nice-sounding phrase identified. The mission should communicate very clearly what the organization is about.

KEY ELEMENTS

In order to focus the organization on the important areas, the leadership should identify key elements when significant effort is needed to accomplish the vision of the organization. From the analysis that has been performed thus far, a number of key vision elements can be identified. These are rather large areas of emphasis. The idea is that no single element is sufficient to capture all of the important areas. Kaplan and Norton called this a Balanced Scorecard approach. Their four basic areas are:

- Customer
- Financial
- Internal Business Process
- Learning and Growth

For many organizations there are other areas that are added to these basic four. In all cases, the key element areas should be an interlocking group that reinforce each other, not a set of conflicting or opposing objectives. Some common additions are community relations, safety, environment, and supply chain.

We are not prescribing any for your organization, but suggest that you consider these and any others that are appropriate. One strong recommendation is that there be no more than nine key elements. Essentially, no organization can maintain a focus on more key elements than that. If the elements are developed correctly most organizations can encompass the important relevant general areas in nine or fewer key elements.

OBJECTIVES

Within each of the key elements, there will be one or more objective statements. For example, if one of the key elements is safety, there may be an objective statement: "Our employees will enjoy an injury-free work environment." Some of the key elements may require more than one objective statement to clearly communicate the intent.

When considering all of the objective statements from all of the key elements, there should not be any conflict or contradictions. When done properly, these should synergistically combine to help the organization perform its mission and accomplish its vision. They should help to weave together the fabric of what the organization considers important and what the organization does. To deploy this idea through the organization, the goals from the higher level become the objective of the next lower level. Plans become the goals, and then each level develops plans to achieve its goals. Eventually the daily activities of individuals in the organization are defined. If this is done in a single step with no feedback from the organization, it is almost certainly doomed to failure. That does not mean that there is veto power from within the ranks, but honest constructive suggestions, ideas, modifications, and improvements should be solicited and considered.

Before going any further, look at Figure 6.7, which shows how the objectives, goals, and plans are deployed throughout the organization. Note that from the executive level each vision element generates at least one objective. For each objective there is one or more goals, each of which is built around SMART principles. Finally, there are plans that will help direct activities to accomplish each goal. As the plans are completed, the goal will be reached. With the

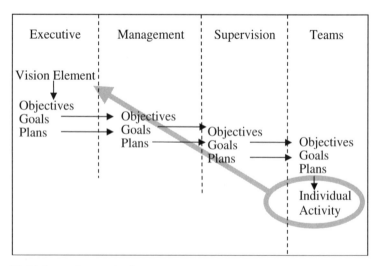

Figure 6.7 Linkage of strategic plan components.

accomplishment of goals, the objectives are met, and as the objectives are attained, the vision element is advanced.

When all of the vision elements are advancing, then the organization will be performing its mission and will be making measurable progress towards its vision. People can understand the relationship of their specific activities with at least one of the vision elements of the organization. If this connection is not clear, it will leave individuals wondering why their work is needed or important.

GOALS

Many people have gone through goal-setting exercises in the past and have failed to see any results from the effort. Only about 25 percent of the population is able to use a goal-setting effort and see any significant difference. This is true even if the goals are specific, measurable, attainable, results-oriented, and time-bound.

"Specific" means exactly that—it should identify the who, what, when, why, where, which, and with whom. Provide enough detail so that there is no inde-cision as to what exactly you should be doing when the time comes to do it. What needs to be done? What results do you want? Who is going to do it? Are there others involved? Where will it be done?

"Measurable" indicates that two different objective people could quantify the progress toward a goal and agree whether it had been completed or not. Tangible evidence is produced based on concrete criteria. How will you know whether the goal has been met? Can the results be quantified? To determine if your goal is measurable, ask questions such as: How much? How many? How will I know when it is accomplished? In general, vari-able measures are better than those that are binomial. Being able to measure that a goal is between 0 and 100 percent complete provides a lot more information than a yes/no approach. At 99 percent complete, a goal is not completed.

"Attainable" means as far as you are concerned. Is the goal possible to achieve? Remember that you can achieve what the mind can conceive. The important thing here is that you believe it can be attained. Goals that are important to you and that you believe in are the ones that tend to be achieved. The unconscious mind can help develop ideas, approaches, and methods when it is a question of something that you truly want. What forces are at play that will help or hinder the accomplishment of the goal? Are there any hindrances that are insurmountable?

"Results-oriented" ensures that something will happen. There is a result anticipated from the setting of the goal. Something is different when the goal has
been accomplished. It may be tangible or intangible. Knowledge is gained,
competency is improved, physical changes are made, and so forth.

"Time-bound" constrains the goal to a period. There is a specific date by which
the goal is to be reached. When should the goal be completed?

Even with SMART goals, only about 25 percent of the population is able to
take a goal-setting effort and use it to advantage. Most people, when it comes
time to review goals, wonder why they were not able to accomplish more. After
all, they had SMART goals, yet nothing happened. In our opinion, the missing
element is a plan of action to accomplish the goal.

One way to develop the action plan for accomplishing goals is to start with the
elimination of obstacles:

- Identify an obstacle to the achievement of the goal.
  - Develop at least three alternate solutions that will address that
    obstacle.
  - Detail actions for each of the solutions.
  - Schedule into a planner, calendar, or personal digital assistant the
    specific tasks by date and time.
- Repeat for every obstacle you can identify.

Most people who meet failure or lack of success with their goal do not spend
the time required in planning. Failing to plan is a plan for failure.

One of the greatest defeats in military history was Napoleon's invasion of
Russia. Over 422,000 troops invaded Russia. Only 100,000 made it to Moscow,
and about 10,000 lived through the retreat back to Poland. In all, 412,000
lives were lost in a failure to adequately account for possible obstacles.
Napoleon failed to consider the possibility of the scorched earth policy of
the Russians and the impact of the terrible cold on his troops. Only summer-
weight uniforms were provided, and temperatures of $-22°F$ ($-30°C$) were
encountered.

As shown in Figure 6.8, the essential parts of the strategic plan can be shown
on just a few sheets of paper. Some have made the effort to get this information
down to a single sheet that is available to every employee. Most will be prettier
than ours; the point is that information in a form that is useful for communication
can be very succinct.

Vision Mission		
Vision Element	Vision Element	Vision Element
Objective:	Objective:	Objective:
Measurements	Measurements	Measurements
Goals and Plans	Goals and Plans	Goals and Plans

• G1
• Plans for
  G1
 	
• G2
• Plans for
  G2
 	
• G3
• Plans for
  G3
 	
•
	• G1
	• Plans for
	  G1
	
	• G2
	• Plans for
	  G2
	
	• G3
	• Plans for
	  G3
	
	•
		• G1
		• Plans for
		  G1
		
		• G2
		• Plans for
		  G2
		
		• G3
		• Plans for
		  G3
		
		•

Figure 6.8 Plans and goals for each vision element.

SIX SIGMA AND STRATEGIC PLANS

In our experience, the odds of a successful Six Sigma implementation can be linked directly to the ties between the Six Sigma implementation and the strategic plan. With the development of a strategic plan, the leadership and managers in an organization are aligned on some common area or vision elements. As the work on developing goals and plans rolls through the organization, it will start to be apparent where there are areas for opportunity. Unless there is already an effort of some sort underway in these major opportunity areas, they become prime candidates for Six Sigma attentions. The Six Sigma process of define, measure, analyze, improve, and control (DMAIC) can be used to understand processes and make the lasting improvements necessary to perform the organization's mission and move towards its vision. Not everything needs to be a Six Sigma project with a full-time Black Belt project leader. Some of the goals and implementation plans are understood well enough that they become part of the daily work of employees. Goals that are large enough to justify a full-time project leader benefit from having a proven methodology, active full-time leadership, appropriate measurements, analysis, and a control system to ensure that the gains are not lost.

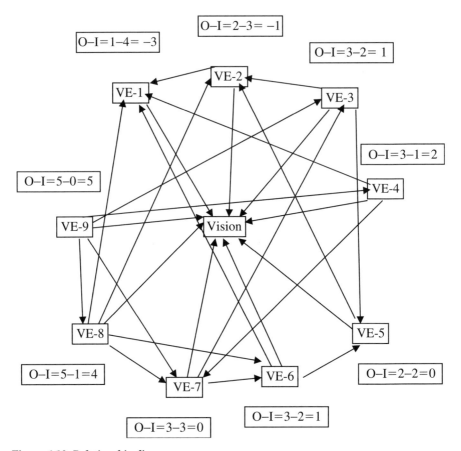

Figure 6.10 Relationship diagram.

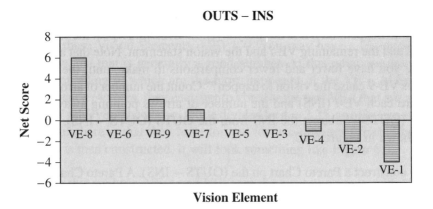

Figure 6.11 Pareto Chart after analysis.

which the VE-*i* show up on the two different charts. Every time there will be a VE-*i* that had very few or no votes that shows up high in the Net Score analysis. The top-vote items will always move down the list, frequently to the last places, as shown here.

In my opinion, the explanation for this phenomenon is that most of us have a good idea of what constitutes the immediate step or two before achieving a result. When asked what is most important, we will tend to select those steps closest to the desired objective.

Let's examine the standard Input Process Output model shown in Figure 6.12. We all tend to focus on the last step of the process that produces the desired output and are quick to identify that last step as the most important. I am not sure we consciously think of it that way, but most, I believe, find it easy to identify that last step and know that if it is not in place the desired outcome won't happen—hence the tendency to think of it as the most important.

When we spend a little time in analysis, we discover that there may be other things that need to be in place first. These show up high in the net score analysis from the relations diagram. Looking at our simple Input Process Output model, as shown in Figure 6.12, these are the inputs and the early steps in the process. When you reduce the variation in the inputs and at the beginning of a process, the benefits carry through the entire process.

We have observed many organizations struggling to make things happen when they establish their priorities based on the votes of knowledgeable people. Progress is a constant battle, and if they ever stop pushing, things seem to get worse than before. The analogy is to pushing a heavy ball up a steep incline. Every bit of progress is difficult and requires constant effort. If the ball is ever left alone and attention turned elsewhere, the ball doesn't just roll back to the starting point at the base of the incline; it actually rolls even farther away, so even more work is required just to get back to the starting point. See Figure 6.13 for an illustration!

This approach is trying to force change. Work is difficult, and progress comes only when there is close attention to detail, usually with close management

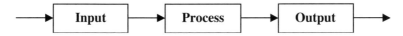

Figure 6.12 Simple process model.

Figure 6.13 The hard way.

scrutiny. We are constantly fighting gravity. It is agony to deal with the burden, and often we cannot see anything else. There is another approach.

Looking at the net scores of "What is the cause?" it becomes apparent that if you can work on the dominant causes, then the other things will result. This is starting with the inputs and the early steps in the process. Using our big ball analogy, this time the big ball is at the top of the incline and we have to get it started down the incline. Once we can get it started, even if we divert our attention to something else for a moment, the ball continues to roll down the incline. Instead of fighting gravity, we now have gravity working for us. See Figure 6.14.

Some get impatient with this approach, since they are not forcing the changes at the last step of the process, and often it feels as if they are not working on the real problem but on something totally unrelated. Here is where many organizations attempting to use Six Sigma start to have problems. They think Six Sigma is a statistical methodology and there must be a total focus on the tools, techniques, and processes involved. This tends to happen in organizations with a preponderance of technical people in the leadership and management roles. While these tools and techniques are necessary and important, and Six Sigma cannot be successfully deployed if they are not in place, they are not sufficient—necessary, but not sufficient. Organizations with this mind-set tend to ignore the people side of strategic plan deployment. Social, cultural, team, and individual improvement within the organization must match the improvements in systems and processes. The people and social issues are important and must be attended to, for successful Six Sigma deployments. A few organizations

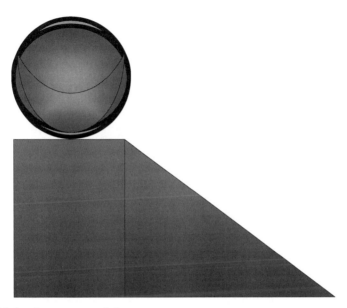

Figure 6.14 Let the process work for you.

attempt to implement Six Sigma while trying to ignore the statistical base of Six Sigma. They give some attention to processes and systems, but do not apply the rigor of statistical analysis of measurement systems and performance. These organizations tend to be led by a predominance of people with nonscientific backgrounds. The people issues must be taken into account, but they are not sufficient—necessary, but not sufficient. Every cycle of improvement needs attention to both areas: the processes and systems (with appropriate statistics) as well as human development. The two are never in perfect balance, but if one is allowed to dominate to the detriment of the other, the organization will not be as effective or as efficient as it could be, and Six Sigma deployment will have problems. If your plans do not include both areas, your organization will be out of balance in very short order, and your Six Sigma projects will not make the kinds of contributions to the bottom line of the income statement that others have enjoyed.

One way to avoid these kinds of problems is to identify the processes that are involved with each of the key vision elements. These processes can be defined at a high level for the executive leadership team. As the processes are identified and then mapped, a number of interactions between important processes will start to appear. A check is in order to ensure that all of the key processes are included at this point. If not, it is likely that there is a key vision element that has been ignored. It is always an enlightening effort to map these processes across the current organizational boundaries. Ripe areas for issues are where processes cross the organizational functions, departments, and so on.

Thus far, we have considered the opportunities that Six Sigma can address, and the strategic plan of the organization. When Six Sigma is used as an implementation methodology for achieving the strategic plan, there is not a deployment problem. Six Sigma is the process that will be used for major projects that are part of the strategic plan.

We see Six Sigma as a tool to help an organization achieve its strategic plan in a more efficient and effective manner. If Six Sigma is not aligned with the strategic objectives of the leadership, it will become an add-on effort that is never embraced by the organization as a way to accomplish meaningful things. Management and Champions must learn their roles and responsibilities in the deployment of Six Sigma. Since Six Sigma is going to be used to accomplish significant objectives and goals that are aligned with the strategic plan, everything possible should be done to stack the deck in favor of success. For something this important—we assume that your strategic plan is important, or you would not have developed it—why leave anything to chance? This is not a test to see whether Six Sigma can work or not—that has been proven in numerous companies—the real question is whether you, the leadership, will do the things necessary to make Six Sigma a success in your organization. Neglecting the necessary foundation for Six Sigma deployments is a sure route to less than desired results.

7

PROJECT SELECTION

Good project selection is a major factor in the early success and long-term acceptance of Six Sigma within the organization. Project selection is a process and, as with any process, if it is not planned and monitored, there is no telling how it will evolve. Every manager, Sponsor, Process Owner, and Black Belt will develop a different project selection process. While we believe some flexibility is needed, unmanaged processes turn into anarchy.

Project identification starts before training of Black Belts. Many organizations get this confused and send people off to be trained as Black Belts before they have projects identified. This hurts in several ways:

- The Black Belt candidates do not have an assigned project in which to apply the skills they learn as they go through the training.
- The number of Black Belts can be totally disproportionate to the available projects.
- Black Belts can finish training or complete a project and not have another project ready.

Doing Black Belt training before projects are identified is the classic "getting the cart before the horse."

We contend that projects are best if they align with the strategic plan and with one or more key objectives from that strategic plan. When projects are aligned with the strategic plan developed by the executive leadership and communicated to the rest of the organization, the project selection process carried out by other managers, Champions, and the Black Belts can be easily checked for alignment. If a project does not align with the strategic plan but has great potential, a case for its inclusion must be made. This may expose areas of omission in the strategic plan that could be addressed in the next cycle. Remember, the strategic plan is a living thing that continues to grow and evolve depending upon internal and external factors. Project selection is an integral part of the deployment process, as shown in Figure 7.1.

Some key elements in the project selection process are:

- Identification of potential projects
- Project criteria
- Project selection
- Project assignment
- Project completion

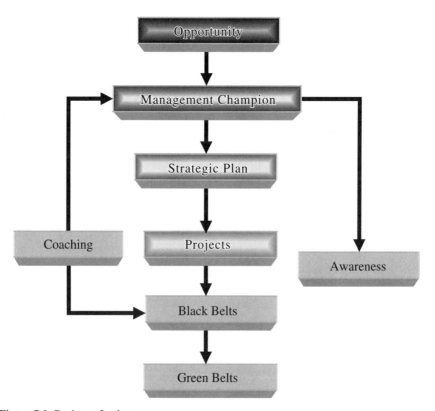

Figure 7.1 Project selection.

- Project evaluation
- Process assessment

IDENTIFICATION OF POTENTIAL PROJECTS

Good ideas are not limited to a few sources in the universe. Potential projects can come from almost anyplace inside or outside of the organization. Many of the sources will not identify a project but rather will describe part of a problem, opportunity, or consequence of a defect. One of the challenges for a Six Sigma company is to have the processes in place to detect these signals in a timely fashion, understand what is being communicated, and then act appropriately.

Customers

Prime beneficiaries of defect reduction are the customers. After all, it is the customer who defines defects in the marketplace. Ignoring customers as one selects projects provides only a self-centered approach to Six Sigma. Many of its benefits can be lost. Jack Welch from GE is on record as stating that GE did not reap the real benefit of Six Sigma until they started focusing on customer issues.

Customers are the people we exist to serve with either products or services. Their ideas and suggestions are a ripe field for the harvest. Unfortunately, the data indicate that only about 4 percent of all customers will ever complain. Complaints are important and should never be ignored—just realize that, without them, you may never hear about some of your problems. For every complaint you do receive, there are about twenty-five more customer problems that exist. Responding to customer complaints is the lowest level of customer interaction. There are a number of other activities that can help you better understand issues from the customer's perspective. Amazingly, one of the most effective is to build a personal relationship and just have a conversation with customers on a regular basis. There are all sorts of statistically valid survey techniques: focus groups, double blind studies, telephone surveys, written surveys, Internet surveys, and so on, and all of these have their place. Not everyone will do them all, but if you do not have an active customer interaction that proactively seeks to identify issues, defects, and opportunities you are missing an essential element of Six Sigma. Do not limit your efforts to the products and services you provide. Sometimes it is the seemingly insignificant things that can be identified as the focus for potential projects. We have done significant amounts of work in the petrochemical business. It is still surprising how difficult it seems to be to match the quantities on a sales order, the shipment, and an invoice to the customer on the first try.

Suppliers

Just as customers provide a wealth of opportunity for identifying potential projects, so can suppliers. Unless there is a trusting relationship between your organization and your suppliers, it may be difficult to obtain useful information from suppliers. This is especially true when the supplier is small and you are a significant part of their business.

Suppliers play an important role in the performance level of your products and services. The overall process is no respecter of the organizational boundaries. It doesn't care if these boundaries are at departments or divisions within the same company or if it crosses several company boundaries. Things can get very interesting when your organization furnishes material or information to a supplier who puts it through their processes and then furnishes it back to your organization. You can become the supplier and customer of your suppliers. Suppliers and customers are important sources of information about many of your support functions. The entire goods/services purchase process can often be enlightening when functions such as purchasing, legal, receiving, transport, accounting, manufacturing, accounts payable, and the like all get involved. While we believe the focus should be on reducing defects to the customers, suppliers should not be overlooked. When one looks at the entire process from beginning to end, customers and suppliers are part of that process. To ignore either as part of the project selection process is a mistake.

Employees

No matter what the level in the organization, there are good ideas and insightful observations that have the potential for fantastic gains. As with customers, the passive suggestion box will generate a very small number of useful leads. Internal focus groups, ongoing metrics, cross-functional discussions, formal process mapping, and so forth are all ripe fields of opportunity. Typically the management has only a high-level understanding of the issues. This 30,000-foot view has its advantages, but suffers from the disadvantage of not being able to see the details. Although the B-52 can deliver very heavy bomb loads from high altitudes, well out of range of antiaircraft fire, the effectiveness can be relatively low. Only when the ground troops are close to the problem and subject themselves to the danger of friendly fire is the precision obtained. Just as the B-52 aircrew is above the conflict and uncertain of the effectiveness or damage from their actions, management can wreak havoc or provide dramatic improvements. The lower-level employees are the ones on the line who can expose the real issues, but like the front-line spotters they run the risk of being destroyed if they or the aircrew (management) make a mistake. Believe it or not, employees instinctively understand this. It can take a long time and a lot of effort to develop the trust and rapport necessary for effective suggestions. One of the more interesting metrics to track is the number of ideas and suggestions that come from below the managerial level in an organization. What fraction results in some positive action? All may not turn into Six Sigma projects, but other continuous improvement efforts help to make this level of the organization a fertile field for ideas.

Extension of Other Projects

It is not unusual for a project team, as part of their work, to come up with other issues or extensions of the current project that might be worthwhile Six Sigma projects. The team does not have to wait until the project is completed to communicate these ideas to the Sponsor and Champion. One caution, however, is to ensure that a new project does not conflict or interfere with the current project.

Competitors

Good ideas and improvements can come from anyplace. Don't be too proud to learn from the competition. While "me too" will never make you the leader, competition may establish new minimum expectations. Consider the vision elements and compare yourself to the competition. Good market intelligence can be very valuable, and don't ignore those competitors who are nontraditional— they may put you out of business. Unless you are leading, changing, and innovating you are behind.

Government and Certification Audits

For some organizations there are regulatory audits and performance reports that can be good leads for areas where opportunities for improvement are hiding. Noncompliance observations from these auditors, if analyzed, can point to system failures. Regulations, laws, and permits all specify what lawmakers consider requirements in the best interest of society as a whole. While some can be capricious or unnecessarily onerous, these laws and regulations do point to directions where business has issues.

Benchmarking

Comparison of your organization's performance to that of world-class organizations can be most informative. Benchmarking is the evaluation of best practices, no matter what the industry, and the development of a plan to adapt the learning to your organization. It is not slavish copying of what someone else is doing. When using benchmarking, be sure you have a defined process and measure the benefits. *Benchmarking: The Search for Industry Best Practices That Lead to Superior Performance* by Robert C. Camp is the book that really started most organizations on serious benchmarking efforts. Suffice it to say that there is a focus on benchmarking processes today, and world-class performance for a specific process is likely found in an industry different from yours. There are groups you can join that do specific benchmarking studies and then report the results back to the participants; consider participating in these studies by furnishing data. Some considerations for benchmarking:

- Know your own performance before talking to others.
- Understand and document your process.
- Be willing to give as well as receive.
- Look at best practices within your own organization.
- Do your research before booking travel. Benchmarking is not industrial tourism.
- Involve process owners.
- Establish that the performance gap is worth the effort.

Benchmarking can also be used on actual projects in the analyze and improve steps of DMAIC. This just demonstrates that Six Sigma application is not necessarily a linear process.

Technology

With the rapid advancements in technical learning and capability, there are new and evolving technologies that may be appropriate for your organization. This does not mean you must be on the "bleeding edge" of every new concept, but that you should have someone looking at new developments in science, technology, and applications. Sometimes it is the new technology that becomes an enabling process for many other processes within your organization. It is instructive to see the difference between the productivity of organizations that have committed to online conferences, a standard desktop with common software, and the like, and those that have allowed the computer and communications technology within their organization to grow without planning or thought.

The Re-World

If you want a quick way to find inefficiency in your organization, do a simple search for the "re-world." With the exceptions of reward and recognition, almost every other word that begins with "re" indicates waste and inefficiency. These are prime places to identify potential projects. Consider the problems where each of the following words (Figure 7.2) might be used.

Unless there is some level of attention to all of these potential sources of projects, it is likely that your organization will miss prime opportunities. We believe that customer focus is the most important, as customers are the ones who decide if we are in business or not.

For a potential project to get on the list for consideration, the ratio-nale for the project must be stated. Project rationale starts with a pur-pose statement or mission statement, which provides the starting point for

Reabsorb	Recapture	Redo	Rehash	Remote	Report	Retell
Reach	Rebound	Redouble	Rehydrate	Remount	Repossess	Retention
React	Recast	Redraw	Reimburse	Remove	Reprehend	Retouch
Reaffirm	Recede	Reduce	Reinforce	Rename	Repress	Retrace
Reallocate	Receivables	Redundant	Reinstate	Renegade	Reprimand	Retreat
Reanalyze	Recession	Reduplicate	Reinvest	Renew	Reprint	Return
Reanimate	Recharge	Reeducate	Reissue	Renounce	Reprisal	Revamp
Reappear	Recheck	Reemphasize	Reiterate	Renovate	Reproach	Revenge
Reassemble	Recidivism	Reenact	Reject	Renumber	Repudiate	Reversal
Reassign	Reckless	Reengage	Re-knit	Reoccupy	Repugnant	Reverse
Reassume	Reclaim	Reentry	Relapse	Reopen	Repulse	Revert
Reassure	Re-coin	Reexamine	Relative	Reorganize	Require	Review
Reave	Reconsider	Refasten	Re-light	Reorient	Reread	Revile
Rebate	Reconstruct	Refill	Relinquish	Repackage	Rerun	Revive
Rebel	Reconvene	Refine	Reliquidate	Repair	Resample	Revoke
Rebind	Reconvert	Reflux	Reload	Reparation	Rescue	Revolt
Rebuff	Recount	Reform	Reluctant	Repave	Reseal	Rewire
Rebuild	Recoup	Refrain	Remake	Repay	Resent	Reword
Rebuke	Recover	Refund	Remand	Repeal	Resign	Rework
Rebut	Recreate	Refuse	Rematch	Repeat	Resist	Rewrite
Re-buy	Recuperate	Refute	Remedial	Repel	Resort	Rezone
Recalcitrant	Recur	Regenerate	Remedy	Repent	Respond	
Recalculate	Red Tape	Regress	Remind	Repercussion	Restitution	
Recall	Redeposit	Regret	Remiss	Replace	Restock	
Recant	Redesign	Regroup	Remnant	Replay	Restore	
Recapitalize	Redevelop	Regurgitate	Remodel	Replenish	Restrain	
	Redirect	Rehabilitate	Remorse	Replica	Restrict	

Figure 7.2 Re-world.

the Champion and Sponsor to explain and coach the Black Belt project leader.

This document creates an overview regarding the project, and includes:

- Description of the issue or concern
- Desired focus of this specific project
- Broad goals or results to be achieved
- Description of the value of this project effort
- Project parameters and expectations, such as resources available to the team and solutions the team cannot consider

PROJECT CRITERIA

As you start to consider projects, we believe that it is much better to spend some time establishing criteria for projects rather than attempting to micromanage every potential project. We recommend that every project be evaluated according to impact or contribution to each of the elements of the vision and each of the four basic areas of the Balanced Scorecard. An excellent project will have a long-lasting positive impact on several of the evaluation criteria and avoid any negative impact. If there are apparent contradictions—for example, a component must be stronger, lighter, and less expensive—be sure these are included in the requirements of the project team charter. If the answer is already known, you do not need a Six Sigma project team; all that is needed is an implementation effort.

Customer Impact

Since these are the folks who determine if we stay in business or not, any impact on present or future customers must be a strong consideration. List all the Customer's Critical Criteria that this project will affect, required and expected. Make an estimate of the current performance and a reasonable estimate of the gain expected for each of the Customer's Critical Criteria, and define the gap for each of the Customer's Critical Criteria.

Rank the project's potential:

1. An incremental improvement in one Customer's Critical Criteria.
2. Small improvements in multiple Customers' Critical Criteria.
3. A significant improvement in one Customer's Critical Criteria.
4. Significant improvements in multiple Customers' Critical Criteria.

5. Will make us best in class or, if not done, will destroy our competitive position for at least one Customer's Critical Criteria.

Financial Impact

Not all of the important numbers are ever known. In fact, Deming was fond of saying that the most important ones are unknown and unknowable. We believe there can be a balance between no evaluation of the financial impact and making financial numbers the only consideration. One of the dangers in this area is giving undue weight to the financial numbers that can be measured. An example is one of our least favorite discount stores. (Since the first draft of this book was written, they have filed for bankruptcy protection.) There is always a shortage of clerks at the checkout lines. Despite the shortage of clerks, when it is time for a break the ones on duty will close down their registers and take a break in the refreshment area, in full view of all the customers waiting in line. When management was asked about the number of clerks on duty, they had a very precise economic formula for calculating the number of clerks needed for the store sales, time of day, and the season. The number of customers who have left filled baskets of merchandise and taken their business elsewhere is unknown and unknowable. Now if this were a one-time event in one store, it would be a special situation, but the experience has been repeated in every one of this company's stores that we have been in, over five different states. While I seldom go in these stores anymore, I always look at the checkout lanes first thing. If there is any line at all, it is time to go someplace else. I am sure not all of the problems of this company are related to frustrated customers leaving because of an inadequate number of checkout clerks, but the mentality and thinking process that allowed this to happen are responsible. The consequences show up across the U.S. as many of these stores have closed and competitors have moved in directly across the street.

Many organizations set a minimum financial contribution for Six Sigma projects. Generally the emphasis is on contribution to profit. For example, if a project frees up warehouse space that is charged at $X per square foot, what fraction can be counted as a contribution to profit? The accepted practice for a Six Sigma project is to count only that which actually adds to the bottom-line profit. In the case of warehouse space that is made available, the amount leased out to someone else after all expenses could be counted. Alternately, if your company leases the warehouse space from someone else, the amount of lease not paid could be counted. Using generally accepted accounting principles, the saving must show up on the bottom line of the income statement.

As another example, a project increases sales by \$Z. The amount claimed by the Six Sigma project is the increased sales less all direct and indirect costs associated with those sales. This is where a reasonable activity-based accounting system will help in evaluating projects. Overhead is frequently allocated on a very arbitrary basis.

Since the late 1980s, many companies have moved to activity-based accounting, which has the advantage of closely matching uses of resources with the products or services that use the resources. Some have been surprised, when they moved from an allocation of overhead to an activity-based accounting approach, to discover that product lines they thought were very profitable were not, and product lines considered unprofitable were actually carrying others. Activity-based accounting can be self-implemented and is not something that requires a massive investment. It can be relatively straightforward, or you can make it very complex with a large infrastructure.

Business Process

Improving the business processes within the organization will add to the bottom line with increased efficiency. Elimination of waste and scrap, and reducing cycle time pay huge dividends. Improved cycle time is an immediate increase in capacity.

Learning and Growth

We contend that the biggest benefit from improved cycle time is the increase in knowledge that results. Every time a process is completed, there is an opportunity for learning. For those willing to collect the data and convert that data into information, there is a learning cycle that also occurs. It does not take very many iterations of learning to outdistance your competitors. A learning organization with improved cycle time for each cycle in learning makes for a very strong competitor.

Time to Complete

Everything seems to take longer than anticipated. As part of the project selection criteria, we recommend that the estimated time be a consideration. Most Six Sigma projects have a targeted time horizon of at least three months and less than one year. That doesn't mean that the project creates delays in order to take three months, if the process can be completed in three weeks. It has been our experience that it usually takes longer in the define and measure stages than most people think. Validation of the project scope and definition, as well as confirming Customer's Critical Criteria, is a longer and more difficult task than most assume. There is always a consideration of the urgency of a specific project. Projects that are so urgent that time cannot be spared to complete the DMAIC process are not good Six Sigma projects. These are usually emergencies that require immediate attention. While some of the tools and techniques from Six Sigma can help even in these times of crisis and fire fighting, they are a poor source of Six Sigma projects. The urgency tends to short-cut the process and the proper evaluation provided by DMAIC. At the other extreme are the issues that are so distant that it really doesn't matter if you address them this year or next. Projects are important and involve some urgency to reach a solution, but they should not require fire fighting.

Risk/Reward

As part of any project, the SWOT analysis form that was used in strategic planning can be used for a specific process. Once a process has been identified with a defined last step and identified first step, it is useful to get the ideas from knowledgeable people about the strengths, weaknesses, opportunities, and threats associated with this process. Frequently, risk is thought of only in financial terms, but there are many more areas to consider, such as confidence in the organization, safety, the environment, reputation, and so forth. Similarly, the reward potential can be in these same areas.

Sequence

There are projects in which a specific order or sequence of project components is important. For very large projects with subprojects, there may be a Master Black Belt for the major project and several Black Belts for smaller subprojects. Even if not coordinated as part of one large project, a particular sequence may be useful.

Organization

Processes cross all organizational boundaries. Often a Six Sigma project is commissioned because the process is unknown or undefined, and it may not be clear who will be involved. Without some consideration of projects in progress or planned, a single part of the organization can become overwhelmed if multiple Six Sigma projects involve a single particular department or function.

PROJECT SELECTION

- View project selection as a process.
- Team members need a good understanding of the organization's customers' business and process, and solid information on how to make improvements on their Customer's Critical Criteria.
- Sources for projects can come from customer feedback, current market conditions, and comparisons with competitors.

As a suggestion, use some of the analysis tools and develop an evaluation matrix. For example, assign relative importance weights to each criterion on a scale of 1 through 7. Then for each project, on a scale of 1 through 5, rank the potential contribution in each criteria area. We did this for the customer impact earlier. You need your own operational definitions for each criterion in advance. For organization, you might want to be sure that there is enough cross-functional involvement to be leveraged but not so much that it is unwieldy. You may decide to assign the weights according to the minimum and maximum number of functions or departments likely to be involved, as in the following example. Your assignments could be totally different from these:

1. The number of departments or functions likely to be involved is unknown.
2. A single department or function, or more than nine departments or functions.
3. Between two and eight departments or functions.
4. Between three and seven departments or functions.
5. Between four and six departments or functions.

Similar operational definitions were developed for impact on each of the other criteria.

We have completed a matrix for a number of projects by multiplying the importance of each criterion by the impact of the project on that criterion and then summing all the products for each project (Figure 7.3).

It is apparent that projects 6, 5, 3, 11, and 14 have the highest scores and are the likely candidates for early assignment. Project 6 is the obvious top candidate. Do not become a slave to the approach; there is not much difference in the scores for 3, 11, and 14. Remember: "Make the tools work for you, don't work for the tools."

Like many tools or techniques, this one can be biased. There is a tendency to oversell the importance of a pet project by giving it an unjustified high impact

	Impact on Customer	Customer	Impact on Financial	Financial	Impact on Process	Process	Impact on Learning	Learning	Time Impact	Time	Risk/Reward Impact	Risk/Reward	Sequence Impact	Sequence	Organizational Impact	Organization	Total
Importance		7		6		4		4		3		5		6		2	
Projects																	
Project 1	5	35	3	18	1	4	4	16	1	3	1	5	2	12	1	24	117
Project 2	2	14	2	12	5	20	2	8	2	6	3	15	2	12	3	24	111
Project 3	4	28	4	24	2	8	3	12	3	9	3	15	3	18	2	36	150
Project 4	1	7	5	30	4	16	4	16	4	12	2	10	2	12	5	24	127
Project 5	3	21	3	18	4	16	4	16	5	15	5	25	3	18	3	36	165
Project 6	5	35	3	18	3	12	5	20	3	9	3	15	5	30	3	60	199
Project 7	3	21	2	12	5	20	2	8	3	9	3	15	3	18	2	36	139
Project 8	2	14	4	24	2	8	2	8	2	6	2	10	2	12	5	24	106
Project 9	2	14	2	12	4	16	1	4	2	6	5	25	4	24	3	48	149
Project 10	5	35	1	6	3	12	2	8	1	3	3	15	1	6	2	12	97
Project 11	3	21	4	24	1	4	3	12	2	6	2	10	4	24	1	48	149
Project 12	2	14	3	18	1	4	1	4	3	9	2	10	3	18	1	36	113
Project 13	4	28	2	12	2	8	2	8	3	9	3	15	2	12	3	24	116
Project 14	1	7	5	30	3	12	3	12	2	6	2	10	4	24	3	48	149
Project 15	4	28	5	30	4	16	4	16	2	6	1	5	2	12	4	24	137
Project 16	3	21	2	12	2	8	2	8	2	6	3	15	1	6	5	12	88

Figure 7.3 Project ranking matrix.

rating. Using this technique will help reduce the bias: Have a group of people develop the importance of each of the criteria and the operational definitions for the project's impact on the criteria. Give the operational definitions of the impact to a second group of people without the ranking of the criteria's importance, and ask them to determine the project's impact on each criterion. Then do the math to calculate the products and sums.

We recommend that the leadership review the criteria at least once per year as part of the strategic planning updates. As part of this review, evaluate the ranking assigned to each criterion and the operational definitions for the impact of a project on each criterion.

PROJECT ASSIGNMENT

Launch a reasonable number of projects. Do not require too many projects, as this reduces the overall benefit of your Six Sigma efforts. Having too many projects is an evil twin to having a project that is too large. Some organizations try to give everyone a Six Sigma project, which causes many different yet equally troubling problems, such as lack of focus and competition for scarce resources. A balance must be struck between the needs of the organization, its resources, and the number of Black Belts. We think the most important consideration is matching the needs of the organization, captured in the project scope, with the skills, talents, and experience of a specific Black Belt. Project assignment is one method to help develop and grow capability among the Black Belts. A well-versed Champion will have some idea of the tools and techniques that will likely be required for projects. By matching this information with the experience and previous projects of a specific Black Belt, the Champion can increase the probability that a Black Belt will have an opportunity to use tools and techniques that may not have been appropriate on previous projects, thus expanding the Black Belt's range of tool use. If you do not keep Black Belts challenged and interested, they are likely to leave for other pastures. While it may seem more efficient to assign a project to a Black Belt who has a proven history of success with that specific kind of project, we suggest that this is a formula to lose Black Belts and gain resistance from new candidates becoming Black Belts.

We recommend that early projects in Six Sigma deployment be very carefully selected and assigned. You want success to demonstrate that Six Sigma works for your organization. The project should be:

- Important to the organization. If there is no value added, who cares? The success of an unimportant project does nothing to help mobilize the organization.
- Reasonably difficult but not impossible. If a project is too simple or has a blindingly obvious solution, success will be met with derision. If the project is impossible, then the murmuring will be that this is another management fad that does not work. Do not try to cure world hunger.
- Capable of being completed in six months plus or minus. Most of us have a limited attention span. If the early projects take too long to complete, Six Sigma will be forgotten in the organization.

PROJECT COMPLETION

Part of the process for project completion includes periodic progress reviews. At a minimum, progress reviews are held with the appropriate Champion. Sponsors, process owners, and Master Black Belts may be included.

DMAIC offers some very logical points for these progress reviews. At the end of each step in DMAIC there is a scheduled review. In addition we recommend smaller monthly progress reports. This allows for easy communication of status, needs, and potential problems.

At the end of the project, a formal presentation is made to the Champion and Sponsor of the project. Others may be part of the group, including other Black Belts, process owners and members of management. We recommend that the project report follow the DMAIC sequence and include the process and tools used at each step. An evaluation of what worked and what did not at each step provides learning for others. A side-by-side comparison of the elements of the original charter and scope of the project with the validated results from the project demonstrates success. It is not unusual for the scope to have been changed and modified during the project. We find it instructive for Sponsors and process owners to see the change from the originally conceived problem statement or project scope to the agreed-upon scope. In general, people who have seen a number of presentations from successful projects learn and are better at developing a properly scoped project.

PROJECT EVALUATION

Some projects have immediate application in other parts of the organization; all that is required is that the information be shared and implementation started. We recommend that even in these applications some effort be made to measure the current level of performance. This allows for unambiguous measurement of the gains, if any. Most of the time we find that there is something a little different that makes it worthwhile to complete the full DMAIC steps, even when it appears that this is a cookie-cutter application of something that has already been learned. Using the learning from a previous project may reduce the time significantly. The small savings in time and effort from using a shortcut process are seldom worth the problems that arise. In our opinion it is much better to go through the full DMAIC steps for every implementation of changes. Learn from the previous projects and speed up without short-cutting the process.

We recommend that as a formal part of the final project report there be a section for identification of new projects that are extensions of the completed project, or come from the investigation of the current project, and are, in the opinion of the Black Belt and the project team, worthy of consideration as additional projects.

Not all projects reach the stated objectives. At this point the Champion, Sponsor, and Black Belt must make a decision on future action. Questions to consider are:

- Is it worthwhile to maintain the current team and make another pass through MAIC?
- Should a new team be created to pick up at this point?
- Should the project be changed in scope?
- Are there really multiple projects hidden within this one project?
- Is it time to accept the gains and move to something else?

PROCESS ASSESSMENT

Project selection is a process, and like any process, it needs an owner. We recommend that this be a member of management or an experienced Champion. An interesting statistic to maintain is the source of ideas for projects that actually are assigned to Black Belts.

Each of us has had experiences of serving on teams where there was poor project selection and a lack of detailed identification. The results came much more slowly and usually caused the team great amounts of frustration.

As you start preparing for Six Sigma Black Belt projects, realize that there are no magic bullets. This is work and effort applied in a planned and intelligent fashion. Success demands more! When projects are completed, one at a time, there is an additive effect. Improvements will tend to be linear with each project that is completed. Exponential growth is possible, and the project "pace" can increase as both management and Black Belts gain more project experience and success.

A truly successful Six Sigma project effort shares the success and challenges. This not only gives encouragement to others, it also allows people with success stories to receive some of the recognition they deserve. Every project is a learning experience. We forge interactive bonds with all our project stakeholders. Leadership, Champions, Six Sigma Black Belts, and employees must all continue to learn and grow. Believe it and achieve it! Doubt it and live without it!

8

COACHING

Recently one of the authors of this book and his wife were on holiday at a resort in a neighboring state. They desired to extend the trip, but their hotel was booked for the remaining week. Their dilemma was whether to return home or to search for a vacancy at another hotel, which was not likely to be available, given the season. They received great news from hotel management on the final day of their stay. The hotel chain was building a new facility two blocks away, and management invited them to stay in what was termed a "soft opening." A complimentary room and meals in a new hotel, which would not be open to the public for a couple more months—how excited they both were! They eagerly listened as the guest services host telephoned to confirm the reservation. She was very courteous and professional in her approach to a fellow employee, and she made these folks feel intensely special. Moments later, excited over their good fortune, they arrived at the front desk of the new hotel. Excitement turned quickly to frustration and embarrassment as the newly appointed front desk manager seemed to lose control at one of those moments of truth when the business's stakeholders come face to face with a customer. It seems this manager was upset with the guest services employee, as she had assigned a room number along with the reservation. This angered the new manager to such a point that she chastised her fellow employee over this seemingly minor incident. It took another ten minutes for check-in due to some sort of computer procedure problems. Alongside the new manager were three employees attempting to assist in the check-in process. The new manager verbally lashed out at these stakeholders for their incompetence at following procedures, and at this point she took full control of the activities. Needless to say, the stakeholders were embarrassed, and so were the new customers. The stakeholders performed no other services, only standing close by the manager with half-smiles on their faces. The customers felt sympathy for these employees. Check-in finally completed, the new manager stepped from behind the desk to escort the new customers around the premises, showing and telling about the hotel's construction and future plans. Her final act was to provide a customer survey, asking them to participate by completing the document and mailing it to the executive manager of the new hotel. The new customers could hardly wait to get to their room. On the survey, they expressed their concerns regarding the need to coach this newly appointed manager, before she further demoralized fellow employees and turned future customers away. This incident serves to remind us that the effort put into coaching pays large dividends. Effective coaching assists in building individual and team confidence, commitment, and the technical and behavioral skills that Black Belts need to complete unexpected, as well as expected, improvement project objectives.

What is coaching? It is the process of instructing, directing, prompting, modeling, and guiding others as they work towards the business's desired outcomes. Coaching is a dynamic and participative process. The amount and type of coaching that individuals need vary, depending on the knowledge, skill, confidence, commitment, and maturity level of the person being coached. Within Six Sigma deployment there are numerous opportunities for coaching, including the coaching of the management and Champions as they develop Six Sigma deployment for the organization, and the coaching of the Black Belts as they lead their projects. As shown in Figure 8.1, these two forms of coaching are often external to the "normal" organizational reporting structure. Of course within Six Sigma there will be coaching of project team members as well.

Dr. Ken Blanchard, in his book *Situational Leadership*, addresses four stages of leadership coaching and participation. Stage one is directing, followed by coaching, delegating, and finally supporting.

In the first stage, coaches direct or instruct individuals having little working knowledge of what to do, or how to do something new. The typical Six Sigma Black Belt has received approximately 160 hours of advanced statistical, leadership, team-building, project management, and people skills training. During the directing stage, it is likely that both coach and Black Belt are in the formative stage of growth. In the formative stage of growth, we are said to be dependent on others for our learning about the details of our duties, responsibilities, processes, procedures, and policies, and about the expectations of the executive management team. Once the coach is satisfied with the Black Belt's level of technical, advanced statistical, leadership, team-building, project management,

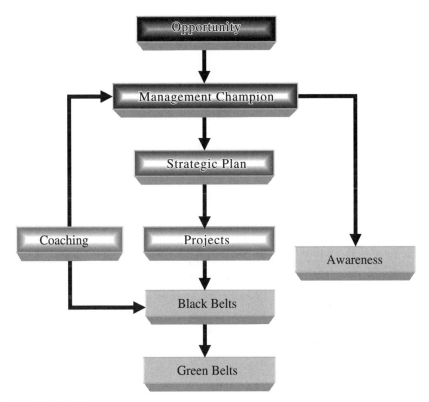

Figure 8.1 Coaching.

and people skills, they move on to the next level of situational leadership, which is called coaching.

In the coaching stage, the Champion is likely to remind the Black Belt of the correct or expected level of performance skills required of a Black Belt project management leader. Successfully completing this stage, our Black Belt moves into the delegation phase of situational leadership and performance coaching. At this point the Black Belt has probably developed into the normative phase of growth, and is said to be independently performing as expected. The Black Belt has most likely mastered the advanced statistical, leadership, team-building, and project management and people skills. However, he or she may be uncertain of the sequence of actions or tools needed in certain situations. This again requires the coach to assist in determining the best possible next steps. During this stage both Champion and Black Belt are likely to be gaining greater confidence and ability to successfully accomplish the organization's strategic plans, goals, and objectives. The final stage of situational leadership and performance coaching is supporting. At this stage, the Black Belt is given an assignment, and the coach is available as needed. This is a stage of inter-dependence, of collaborative team efforts that are designed to assist the Black Belt in successfully discharging all project deliverables. Now, the Black Belt has doubtless earned the trust and respect of all organizational stakeholders as a collaborator and as a true professional. There is conceivably less time spent between coach and Black Belt during the supportive stage of growth.

Coaching is a normal part of the work relationships between the businesses executive management team and Champions, as well as Champions and Black Belts. Coaching is proactively influencing the performance of others. It often determines whether the individual uses a particular skill set, or attempts to achieve the desired business results. This is especially true regarding participants applying recently acquired Six Sigma methodologies.

Coaching builds positive affirmations, commitment, motivation, and skills to handle a situation or manage interpersonal relationships successfully, and to ensure the consistency of Six Sigma methodologies quality. Adding to another person's emotional bank account, avoiding the use of generalities, labels, or assumptions, and actively practicing the skills of versatility will greatly enhance the coaching experience for all concerned. Coaching is a value-added productivity investment for all organizational stakeholders and an activity not to be taken lightly.

Many businesses would like to improve their people, processes, and systems to attain greater market share and profitability. Some businesses are behind the technology curve, and do not invest in training and coaching their employees because they can ill afford the loss in productivity, or the extreme cost factors often associated with traditional training and coaching processes. Some organizations fail to recognize that training and coaching their employees builds a competitive business advantage that contributes to satisfied customers, fewer mistakes, less rework, and greater profits to the bottom line.

We believe that excellence, making changes, and confident expectation are key to an organization's ability to achieve world-class performance. Excellence requires continuous improvements, benchmarking, the constant pursuit of excellence in products/services, and the capability to do things right the first time. Making changes is the way organizations gain competitive advantage. Coupled with excellence, making changes becomes part of a powerhouse combination for achieving world-class performance, which contributes to improved profitability. Confident expectation provides information about the future, which allows organizations to be in the right place at the right time, with their excellent and innovative products/services ready for market.

This triad (excellence, making changes, and confident expectation) arms organizations with the competency to predict their customers' future needs and to develop excellent products and services to meet them. We believe that these attributes of world-class performance will provide survival skills and add to an organization's ability to thrive, even in times of chaos.

Coaches assist in identifying the performer's specific needs, objectives, and learning outcomes, based on the organization's strategic improvement opportunities and desired business results. Coaches partner with Black Belts to accomplish desired business results by delivering specific training products and services within time and cost constraints. Coaches focus on the Black Belts' concerns about identifying the personal and business improvement opportunities of the targeted participant population and assist in discovering timely and cost-effective solutions to accomplish the organization's business objectives.

Coaches assist Black Belt partners in benchmarking current issues and concerns, pressing goal achievement into play, and creating a road map to success. Coaches must assist in mapping pathways showing how to get from the present to the future. Coaches assist partners in recognizing that participants' values and beliefs often require personal awareness training and coaching to move individuals towards the organization's desired business results.

Coach's focus on fulfilling a Black Belt's need to become more productive, get into and out of training quickly, and become more confidently responsive to the organization's needs. Coaches help stakeholders to experience less painful training, offering a wide variety of delivery systems, depending upon what is most cost-effective and produces the desired competence. Coaches are personally committed to their Black Belt partners' return on investment, to their successes and growth, and to developing long-term sustained success.

Six Sigma training draws on the concept that for any endeavor you must start with the end in mind. We consider world-class performance, as defined by the customer, to be the objective that all businesses seek to achieve. The overall business strategy must be developed to focus efforts towards successful achievement of world-class performance. All stakeholders need to be able to make a direct link between the daily activities they are performing and at least one of the six to nine essential performance areas of the business strategy.

Improvements come in two different ways. The first approach is to improve the business systems and processes, using all of the process improvement tools.

The second is to develop the capability of people as individuals, in teams, and as an organization. Efforts to improve one approach to the exclusion of the other will not be as effective as will an effort that considers and includes both.

The result of successful coaching is the melding of the business focus through strategy. Coaching then binds together the synergy of people and processes aimed at world-class performance, in the areas that are critical to the organization's customers. The process does not care where the organizational boundaries lie. Boundaries can be in one company or department, or in totally different companies. If you were to purchase one of your suppliers, would that make any difference to the process? If nothing changes, it will not matter if the supplier is part of your company or organization or comes from the outside. The process crosses organizational boundaries.

The better each step in the process understands the ultimate Customer's Critical Criteria, the more likely you are to make the changes necessary to meet them. We understand that organizations hire people to produce results, not to do tasks. Obviously, however, we must perform tasks to produce results. The key for coaching, then, is to work on the performances and behaviors of the tasks that produce the desired business results.

We feel that when training or coaching needs are identified, they should be scrutinized to expose the specific components of knowledge and skills that will correct the need or create the desired performance or behavior. Success of our internal customers is a prime focus in all we do, in order to reach the organization's desired business results.

Training is hardly ever a solution by itself. Rather, it is coupled with coaching and integrated into an organizational environment where expectations are clearly defined and measured, and its relationship with all other aspects of the work environment are understood and leveraged. This is important because, at those decisive moments when there is no one to ask and your employees need to act, they will either help or hinder the achievement of your organization's desired business results.

THE PURPOSE AND PROCESS OF COACHING

The purpose of coaching is to plan and design a comprehensive curriculum that identifies and prioritizes all the specific significant knowledge and skills required by a given segment of the organization. This is accomplished through a process in which coaches help Black Belts to understand (1) that the organization's vision is to delight the customers, (2) that the organization's values are those of customer focus, development of people, building stakeholder partnerships, improving quality, and building teamwork, and (3) that measurements are customer-focused by utilization of Customer's Critical Criteria, cost reduction, profitability increases, and meeting the organization's desired business results.

Measurements include a people-focus approach, using the organization's strategic plans, training, building of morale, and Six Sigma tools and concepts, as well as coaching. Measurements also include the organization's internal focus regarding productivity, process compliance, cycle time reduction, reject rework rates, cost of quality, total quality management, Six Sigma, and world-class performance measurements. Finally, measurements include the organization's supplier focus regarding on-time delivery, cycle time reduction, total quality management, and Six Sigma partnerships.

Six Sigma Black Belts need coaching regarding the organization's strategic plans, Six Sigma implementation, mistake proofing, and overall process effectiveness. Black Belts need coaching in the art of process mapping, project management, quality functional deployment, management and leadership skills, activity-based management, activity-based costing, and proper coaching techniques.

Black Belts need coaching in concepts of the organization's specifics regarding Customer's Critical Criteria, in concepts of the process as a customer, in the concept that quality is critical, and in the concepts of defect recurrence prevention, speaking with data, and strategic management.

Black Belts need coaching in the principles of strategic plans, people empowerment, and involvement. They need coaching on the principles of processes and results. They need coaching on the principles of total systems, decisive moments, and world-class performance. Coaching is not designed to take on the performance activities that rightfully belong to the performer in question. It is a process of assisting the performer to learn to solve problems, take responsibility for decision making, and improve his or her personal skill sets for managing interpersonal relationships.

Coaches gain active participation in the process of attaining desired outcomes through asking open-ended questions. For example, a coach might ask, "How do you plan to handle the data-reporting process with the new executive management team member?" The coach might then ask the performer, "What reactions might you expect due to the technical nature of the report? What will you do or say if such-and-such occurs?" Each of these open-ended questions requires the performer to give more than a simple yes or no answer; thus, the process solicits the performer's active participation.

A discussion of the pros and cons of each approach or suggested solution will help the performer arrive at a resolution that can be agreed upon. Coaches need to be skilled in root cause analysis techniques, anticipating problems or barriers that might arise as a result of selecting a particular solution. Coaching is the enhancement of the performer's self-esteem. Positive affirmations are the order of the day; accentuating the positives and eliminating the negatives assists the coach in enhancing another person's self-esteem. Building on collaboration, synergy, and innovation contributes to both motivation and personal commitment regarding the actions agreed upon as solutions to the situation.

Next the coach reviews or summarizes the responsibilities and action steps to be completed, so both parties can confirm commitment. This is the final check for understanding in the coaching process. Expressing confidence in the performer—letting the performer know you feel that he or she will successfully accomplish the objectives—is yet another motivational factor to be exercised in this final coaching step. Use specifics and be sincere in your delivery. Never blow smoke—the performer will most likely recognize it for what it is and lose confidence in your coaching. Finally, end the discussion on a positive note, and set follow-up meetings as necessary.

PAYOFF OF COACHING

Coaching is conviction-driven and should never compromise one's beliefs. Coaching is the art of helping the performer practice becoming a world-class performer and to accept change when and as needed. Coaching is having the ability to respond consistently and predictably. Coaching is based on honesty, trust, and integrity. Having the ability to understand another person is the heart of coaching, and it is something each of us can do for another person. Helping others to see themselves, to gain new insights and skills, to gain maturity, and to learn the fine art of balancing their life experiences is one of the greatest gifts to be gained by both the coach and the performer.

The coaching process is complete when you have organized results of the studies and have mapped out a comprehensive plan. The plan identifies and prioritizes all significant training needed by the Black Belt reviewed or studied. It includes goals and objectives for both the coach and the Black Belt. Coaching is performed when there is a problem to be resolved, when activities are not fully met according to the parameters of the project plan.

We coach when there is an opportunity for action that is required for effectiveness or efficiency, or when we acquire a new set of skills. Coaches must fully understand and be able to use the skills or specific actions they require or desire of a Black Belt. Taking time to acquire these actions or skills training adds greatly to the coach's ability and success. We strongly suggest that coaches of Six Sigma Black Belts attend the Green Belt training to refine their understanding of the Six Sigma methodologies.

We feel that effective coaching is a behavioral process that involves telling stakeholders how to do something, showing them how it is done, and then letting them demonstrate competency in that performance. Coaches then give praise for things completed correctly or nearly correctly. Appropriate feedback and redirection occur regarding things not completed as required.

Reprimands do not occur during the training process, as this tends to stifle and demoralize the performer. Once Black Belts have successfully demonstrated acceptable performance, they are determined to be competent. Coaches can now begin the process of correcting by reprimand, if the need arises. Seldom if ever will a Black Belt have performance problems this severe. We have seen the need on occasion, when there are team members who are not performing.

Reprimands follow the process of first trying to understand another person's actions and motives. We do this by first asking why. Next, we specifically inform the performer why the action or behavior concerns us. We do not use labels, generalities, or assumptions, but rather we pinpoint the concerns specifically, in such a manner that they can be seen, heard, and otherwise measured by others. Next, the coach asks the performer how he or she intends to correct this defect. It is vital, to both coach and performer, that these ideas come from and are commitments of the performer. We are, after all, seeking improvement, which is more likely to come from personal commitment than from compliance with someone else's ideas. Improvement of personal performance is most likely to come from the recognition of the consequences that the performer is producing from either the desired or the undesired behavior.

Understanding something about antecedents, behaviors, and consequences of performance and behavior will greatly assist coaches in effectively determining when to support or correct the behaviors of those they are coaching. Antecedents are the events or conditions that precede a behavior. Consequences are the rewards gained by or denied to the performer following a behavior. When examining consequences, examine both desired and undesired consequences of the performer's behavior.

Let's use a Six Sigma Green Belt as an example of how this antecedent, behavior, and consequence process can work. The antecedent is formal Six Sigma Green Belt training and certification. The desired behaviors would include periodically collecting, recording, documenting, and analyzing raw data. The behavior is the action exhibited towards the desired performance. If our Green Belt performed as expected, what would be the consequences of his or her actions? Listing each consequence specifically would likely help us to see there are both positive and negative consequences to this desired behavior. For each consequence listed, the coach determines if it is personal to the performer or to others, and if it is certain or doubtful as an impact upon the performer. Finally, determine if it is immediate or delayed as an impact on the performer. Add up all consequences, identifying them as personal to the performer, having a certain impact on the performer, and having immediate impact on the performer. This identifies this consequence as highly likely to be repeated. The performer is gaining personal, certain-to-self, and immediate gratification for this action.

This process helps identify the consequences that are motivating the performer to take the action. Keep in mind that these may be positive and or negative consequences. The coach examines both desired behavior and undesired behavior, to obtain a more complete picture of performance motivators for specific and pinpointed performance concerns. Once these are identified, measured, and analyzed, the coach can then set about to correct each in turn. It is of interest and importance to note that consequences of performance that are determined to be of a personal nature are certain and immediate to the performer, and are likely to be repeated more than 80 percent of the time. Armed with this knowledge, coaches can predict the performance of individuals with some degree of accuracy. This in turn helps Champions develop better planning, delivery, coaching, and rewards for desired behaviors and performance. Keep in mind that Black Belts are selected form the organization's top technical arenas; they possess many management, team-building, technical, and people-related skills. Black Belts are more likely to seek information, ideas, and effective collaboration regarding their responsibilities than to need coaching activities. These activities suggest that Black Belts are more likely to build commitments as they draw on the knowledge, skills, synergy, and collaboration of the other stakeholders assisting quality project improvement processes.

Coaching begins with asking open-ended questions, followed by effective listening for message content and associated emotions. Content gives the coach clues as to the issues, concern, or details of information they are seeking. The emotional content can provide clues as to the participant's frustration, fear, anxiety, excitement, happiness, sadness, and so on, regarding the concern being shared with the coach. Listening can be difficult; often we interfere with the message being sent by asking questions inappropriately, offering opinions, making suggestions, or commenting in general.

As a listener, attempt to understand the message being relayed to you. Once the speaker has completed establishing the purpose and importance of this communication, clarify the details. This is best practiced by paraphrasing or quoting the message content as you understood it, and folding in the emotion as you felt, heard, or witnessed it. This then gives the speaker an opportunity to correct your understanding of the specific points of the intended message. Information gathered to this point will have significant impact on the remainder of the coaching session. The coach continues to seek input, using open-ended questions, until both are satisfied that the intended message has been correctly understood. Coaches might need to assist in identifying and discussing constraints or barriers viewed as impediments to the successful resolution of the situation at hand.

Listening first without interruption and then responding with empathy, as opposed to sympathy, is another critical coaching skill. Empathy says that you identify with and understand another person's situation, feelings, and/or emotions. Sympathy is the act of sharing another's feeling by agreement, compassion, loyalty, devotion, and allegiance in a relationship in which whatever affects one person correspondingly affects the other.

Next, the coach and Black Belt collaborate in reaching the desired all-win solution and commitment to corrective actions required by this issue. Asking what one desires to accomplish begins the discovery process for gaining effective and desired solutions. Coaches do not take on performance activities that rightfully belong to the employee. It is the coach's duty to remove barriers to desired performance, to assist in solving problems, and to assist others in improving desired performance skill sets, as well as successfully managing interpersonal relationships.

Stress is a reality of both coaching and of the activities required of Six Sigma Black Belts. Coaches actually work in the presence of the greatest stress producers known—other people. Few coaches responding to the needs of others escape emotional involvement, as they generally feel deeply about efforts to resolve the problems presented to them. It is best not to possess but rather to assist others, as they hold their own visions, values, attitudes, and paradigms. Let them make their own conscious efforts to form behavioral responses regarding the consequences of desired performance outcomes. Coaches would do well to reduce or eliminate their own emotional involvement with participants and act more as counselors.

Black Belt performers remain the vital variable in the coaching process. Their perceptions are reality to them; their reactions to these perceptions are central and powerful factors in the coaching process. It is indispensable to understand both the performer and the problem before attempting resolution. When coaches hear performers say they feel helpless, maybe this feeling is due to the lack of sufficient knowledge or training in a specific area. Pausing, asking for more information, listening for content and emotion, will help you to understand this feeling of helplessness. Listening includes paying attention to the performer's body language, voice modulation, facial expressions, and hand gestures as providers of clues about those we coach.

As coaches, if we are to be effective, we must suspend our own judgments in order to gain trust and vital information from those we coach. The caution here is to delay the attempt to fit the communications we have received into some category based on our personal knowledge and experiences, and then to attempt to announce solutions. Happy coaching!

9

BLACK BELTS

Black Belts are typically full-time project managers. Projects that meet the project selection criteria are matched to the available or potential Black Belts for completion. The Black Belt will lead the work on the project through the major project steps of define, measure, analyze, improve, and control (DMAIC). The Black Belt leads presentations and develops reports following a successful project. These can be made to Sponsors, process owners, other Black Belts, management, and the various departments or functions within the organization.

Drawing on the concept that any endeavor starts with the end in mind, we consider that world-class performance, as defined by the customer, is the objective that all businesses seek to achieve. The organization's overall business strategy must be developed to focus efforts towards successful achievement of world-class performance. All stakeholders need to be able to make a direct link between the daily activities they are performing and at least one of six to nine essential performance areas of the business strategy.

Improvements come in two different ways:

1. Improve the business systems and processes using all of the process improvement tools.
2. Develop the capability of people as individuals, in teams, and as an organization.

Efforts to improve one to the exclusion of the other will not be as effective as will an approach that considers and includes both. The result of a successful Six Sigma deployment application is the melding of the organization's focus, through strategy and through binding the synergy of people and processes that are aimed at world-class performance. World-class performance for each organization derives from successfully meeting requirements and expectations in the areas that are critical to the organization's customers.

We believe that Six Sigma deployment requires intensive training of certain people in order to obtain desired business results. Specific Six Sigma training teaches new skills, in a timely and cost-effective manner, to specific employees in a specific job function. We believe that value-added employee development is the process of getting people to effectively utilize and continue to develop skills they have previously learned.

In our opinion Black Belt candidates are the "best and brightest" from the organization. A quote from the 2000 GE annual report reinforces this concept.

> It (Six Sigma) has, in addition to its other benefits, now become the language of leadership. It is a reasonable guess that the next CEO of this Company, decades down the road, is probably a Six Sigma Black Belt or Master Black Belt somewhere in GE right now, or on the verge of being offered—as all our early-career (three to five years) top 20% performers will be—a two-to-three-year Black Belt assignment. The generic nature of a Black Belt assignment, in addition to its rigorous process discipline and relentless customer focus, makes Six Sigma the perfect training for growing 21st century GE leadership.

Most Black Belt candidates have at least an undergraduate degree and some comfort in dealing with mathematics. It is not necessary that they be well schooled in statistics, but the position does require that they be willing to learn how to apply statistical techniques. While Black Belts will learn to apply statistics to solve problems, the statistics are just tools. We are firm believers that the tools work for you, not that you work for the tools. To that end, each Black Belt needs a good all-purpose statistical software package. There are several on the market that are more than adequate. We recommend that you select one and use it across the entire organization. Depending on your willingness to invest in software, there are a number of specialty programs that do a few things, generally better than the all-purpose packages. Design of experiments, failure mode and effect analysis (FMEA), reliability engineering, TRIZ, and project management are a few of the areas where specialty programs can be useful.

With the concentration on statistics required, Black Belts are frequently people with at least a B.S. degree in science, engineering, business, math, statistics, or the like, though successful Black Belts also come from outside these disciplines. Black Belts are usually not new employees but rather those who have several years of experience within the business. Often, people selected are targeted as future leaders in the organization, and Black Belt training and experience becomes a grooming step.

In short, your Black Belts are the best and the brightest. Organizations that have successful Six Sigma efforts make Black Belt selection an important activity. Some organizations have announced that stakeholders seeking a management position must first become Black Belt qualified. Consider the power brought to the management of your organization if all of the new managers moving up through the ranks understand and have personally led several Six Sigma projects. The management view of the entire organization will start to change, along with the understanding of what could be and what it takes to make change happen.

There will be a need for other team members that the Black Belt must recruit to work on the project on a part-time basis. This requires that the Black Belt negotiate successfully with the appropriate management and supervision to get the help needed. In addition, the Black Belt will need to convince the appropriate individuals to work on the project. We recommend that these new team members be given additional training to a Green Belt level of competence. Black Belts must assist the Green Belt team stakeholders by identifying their specific needs, objectives, and learning outcomes, based on the organization's strategic improvement opportunities and desired business results. Black Belts assist Green Belt stakeholders by delivering specific training products and services within the desired time and cost constraints of the organization.

Each individual of Six Sigma organizations must be personally committed to their return on investment, their successes, their growths, and to developing internal and external long-term loyal customer relationships. This is especially true of the Black Belts and people serving on Six Sigma project teams. We understand that organizations hire people to produce results, not simply to perform tasks. Obviously, employees must perform tasks to produce results. The key, then, is to work on developing effective behaviors and performance of the tasks that produce the desired business results.

SELECTING BLACK BELTS

In our deployment model, there are a number of activities and steps that take place before Black Belts are needed. First, the senior leadership must understand the opportunity for your organization, and a decision must be made to use Six Sigma as a methodology to achieve much of that opportunity. The strategic plan with vision elements must be clearly communicated and understood. Objectives, goals, and plans are needed for each vision element. Management, Champions, and other members of the leadership of the organization need to understand their responsibilities and roles in the deployment of Six Sigma. A project selection process needs to be in place with an owner of the process identified. Before Black Belts are needed, there must be an inventory of projects ready to be assigned to Black Belt candidates. Only then is there a need for Black Belts. Thus far the deployment plan has a very linear approach (Figure 9.1). Steps are defined in order. It is unlikely that many organizations will move through deployment in such a fashion. Also there are the inefficiencies of the re-world: Strategic plans are reviewed and modified. Managers are reassigned or replaced, each necessitating some effort to ensure that people are cognizant of their roles and responsibilities.

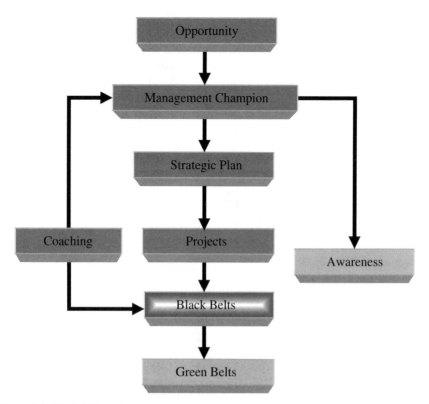

Figure 9.1 Black Belt project managers.

As you prepare to select Black Belts, we suggest you consider at a minimum the following areas:

- Your organization's criteria for Black Belt candidates
- Number of Black Belts
- Duration of assignment
- Future career path
- Organizational reporting
- Compensation

Criteria for Black Belt Candidates

- Rating/Performance. Do not select the people that no one else wants. Very quickly you will have as Black Belts people who are not wanted in the organization. At the end of their assignments, it will be difficult to find them meaningful assignments.
- Management potential. We believe that most Black Belts will be on the list for future management assignments. For some, leading a Six Sigma project team may be their first experience leading people.
- Technical capability. In our opinion, candidates will have demonstrated the ability to master the technical requirements of their previous assignments.
- Experience. Usually a minimum of three to five years. New hires don't know their way around the organization yet and are still learning how to work in the organization, in addition to learning to lead Six Sigma project teams.
- Functional knowledge/Competence. While this requirement is similar to technical capability, Black Belt candidates also understand how to work within a specific department or function.
- Self-starter/Ambition. This is difficult to teach. Select the people who have initiative.
- Respect within the organization. In any organization there are those who are respected, valued, and wanted, and those who are not—select the former.
- Problem-solving ability. People who enjoy solving problems will do better than those who do not.
- Process orientation. People can be taught how to think about processes. Those who already have this orientation are further along the path.
- Sense of urgency. The faster you make the changes in your organization, the sooner you will realize the benefits. Select people who understand that speed without taking shortcuts in the Six Sigma process of DMAIC will return dollars to the bottom line faster.

- Interest in being a Black Belt. People who do not want to be Black Belts will probably go through the motions, but their hearts will not be in it. Management can help create the desire within the organization through the way they treat Black Belts and the assignments Black Belts are given.

Number of Black Belts

A common question is: How many Black Belts do we need? We contend that this is the wrong question to ask. Instead, answer the following questions, which will lead to the answer of how many Black Belts are needed.

- How fast do you want to improve?
- How many people can you devote to improvements?
- Is your organization ready for Six Sigma?
- How many teams can be supported?
- How many qualified candidates do you have for Black Belts?
- Has management done its work in identifying potential projects?

We have given you a few things to consider. Once your company is committed to Six Sigma implementation, it is essential that there be enough projects identified to keep all of the Black Belts adequately challenged. It is a terrible waste of resources to have qualified Black Belts ready to tackle important issues when there are no projects for them. This generally means that the project selection process is not functioning properly. Highly successful organizations that have implemented Six Sigma generally have more projects than Black Belts.

There are a number of ways to estimate the number of Black Belts needed. This is a management decision. The following section lists three approaches. We recommend that you apply all three, to see if you get widely different numbers. If so, some of the management work needed for Six Sigma deployment has probably not been completed.

Approach 1: Estimate from Waste in the Organization

Estimate the current performance level of the organization. If you do not know, use three sigma, which translates to ~25 percent of sales as waste. Because Six Sigma processes have less than 5 percent of sales as waste, the difference is 20 percent of sales (Figure 9.2). Assume that half of that, or 10 percent, will be eliminated, and allow a five-year implementation plan. Decide on the value expected from each project (typically $100,000/yr +) and the

Cost of Poor Quality
% of Revenue

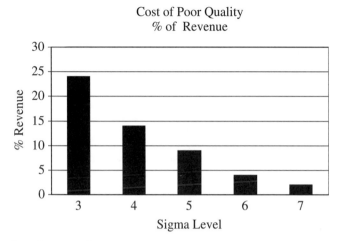

Figure 9.2 Cost of poor quality.

number of projects per Black Belt (usually three to four per year). Then do the math:

Number of BB = [(sales) × (0.2) × (0.5)]/

[(5) × (project value) × (projects/year/BB)]

Approach 2: Calculate Number of Black Belts as a Percentage of the Employee Population

Black Belts should be 0.5 percent to 4.0 percent of the employee base: 0.5 percent with a high concentration of unskilled labor, and 4.0 percent for those with very fast implementation plans.

A caution: If you have too many projects, they can overlap and interfere with each other. With this concentration, there will probably be a number of big projects, each led by a Master Black Belt, in which Black Belts will have part of the overall larger project.

Approach 3: Determine the Number of Black Belts Needed to Meet Project Needs

With this approach, the leadership defines a number of projects in advance. Each project is estimated for the amount of time required to complete. Based on the projects and the amount of time, only enough Black Belts are trained to meet those needs. On an annual basis, new projects are evaluated, and the need for additional Black Belts is determined.

Sensitivity Analysis

We suggest that for any of these approaches you do some sensitivity analysis to determine the impact of the assumptions. This includes actions and considerations such as:

- Vary the estimates of waste in the current organization from, say, three sigma to four sigma.
- Average project returns may be from $100,000 to $1,000,000.
- Implementation time could vary from two years to seven years.
- Recovery of the estimated waste may be from 50 percent to 90 percent.
- Average time to complete a project could vary from three months to a year.
- Identified projects may fail or turn into multiple projects, so vary the number of projects from 50 percent of the estimate to 125 percent.

Duration of Assignment

How long will Black Belts be project leaders? If Black Belt becomes a permanent assignment, the best and brightest will avoid the development assignment. Also, if Black Belts are not moved back into the line organization as leaders, much of the leverage is lost. We believe the following should be considerations when determining how long a Black Belt stays in that role.

- The assignment should be long enough for the organization to realize the benefits from having invested the training in the Black Belt.
- The assignment should be short enough that the Black Belt does not get "burned out" working on projects.
- Continue to develop people over time, and each year a larger percentage of the key people in the organization will have had the experience of being a Black Belt.
- What are your training costs for the first wave and then for replacement Black Belts?
- Two to four years seems to be a good balance.

Who is responsible for helping with a Black Belt's future career? If someone is not assigned this responsibility, it will be neglected. We recommend that the Business Manager and the Champion both have a role to play. The Business Manager has all of the organization's Black Belts at the top of the list for any management opening. A simple process is that everyone with two years or more as a Black Belt be considered for any opening. Only if there is overwhelming evidence that none is qualified for the position is anyone else considered.

Champions have the responsibility of ensuring that the qualifications, successes, and ambitions of each Black Belt are communicated effectively to all Business Managers.

Some organizations have a semiformal mentoring program for Black Belts. Someone from outside the business unit where the Black Belt reports meets with the Black Belt on a regular basis to discuss issues, problems, concerns, potential opportunities, and so on. The mentors need to be high enough in the organization to be able to visit with the Business Managers as peers.

Black Belt Future Career Paths

One option is to promote Black Belts to responsible leadership roles. In today's market Black Belts are in high demand, and if you do not take advantage of the talent it may move elsewhere. Some are willing to take on the responsibilities of a Black Belt in order to improve their chances of getting more responsible positions. If you have suggested this in your communications to the organization, be prepared to live up to it. This does not mean that every Black Belt will be promoted to management, but you can be guaranteed that the organization is watching, and if others are promoted over Black Belts, you have clouded the message.

Some Black Belts find that they want to spend a good part of their career leading Six Sigma project teams or coaching other Black Belts. These people can be developed into Master Black Belts. They usually have completed six to ten successful projects and have developed their communications, project management, and statistical tools beyond the level of the normal Black Belt. You will probably not want to have more than one Master Black Belt for every ten or so Black Belts. Err on the side of too few rather than too many Master Black Belts. We feel that Master Black Belts always need to have an active project, even if they are devoting a large portion of their time to coaching other Black Belts.

Those Black Belts who enjoy the work and are content to lead projects can be maintained in the position of Black Belt. There is a danger to this approach, however. If you have too many of these "career Black Belts," there will be fewer opportunities for new people to lead meaningful projects, and you will miss the synergy of integrating successful Black Belts into the rest of the organization in other meaningful jobs.

For those Black Belts who are not ready or do not show the required skills and ability to move into management positions, we recommend a lateral move to a different opportunity. This provides meaningful work for the Black Belt and opens the opportunity for other people to be developed as Black Belts.

The key is to have options under consideration early for each Black Belt. The position of Black Belt is a cherished one, and management cannot allow someone to occupy that position for too long. It is really a development position for other jobs in the organization. If consideration is not given to the Black Belt incumbents and to potential future Black Belts who will benefit from this development, the individuals and the organization all lose.

Organizational Reporting

In our opinion, there are a number of advantages if Black Belts stay in the business unit where they are working. First of all, it keeps the focus on the importance of the business unit's customers. Further, Black Belts will be committed to meeting the needs of the business unit. The potential of having a separate function with its own priorities and objectives is avoided. The objectives and goals of the business unit are linked directly to the Black Belt's projects.

This approach produces less bureaucracy, because it allows you to avoid building a parallel organization that is outside the normal line function. This was a common problem with many TQM implementations. By keeping the Black Belts within the normal line function, the managers of the business unit are held accountable for generating additions to the bottom line. The success of Six Sigma projects becomes an important performance measure for managers. Many successful companies have tied stock options and variable pay to the contribution of successful Six Sigma projects within their areas of responsibility.

Black Belts who remain with one business unit have increased leverage when back in the line organization. As part of the line organization, Black Belts maintain their contacts and associations within the organization. If their next assignment is within the same business unit, they will be more in tune with what is happening and understand the issues better than they would if isolated in a separate "Six Sigma Department." Keeping the Black Belts within the line organization also ensures that they are on the personnel list for the business managers. Pay actions, promotions, EEO calculations, and so on will all have the Black Belts included.

Compensation

Black Belt compensation is an issue that must be addressed by management. If Black Belts are really important project managers and have come from the ranks of potential leaders for the organization, their compensation should reflect this. Candidates are probably already well compensated, given their performance to even be considered as a Black Belt.

Once a project manager is trained and working as a Black Belt, how will the compensation issue be addressed? Several options are common practice, but no one we know does all of these:

- Black Belts receive a boost in base pay when they complete Black Belt training.
- Black Belts are given an additional boost in base pay when they successfully complete a certain minimum number of projects (the number varies).
- Black Belts receive a fraction of the savings from projects they lead.
- Savings for all Black Belt projects are summed over a given time period and a fraction awarded to all Black Belts in the organization active during that time period. All Black Belts receive something.
- Black Belts are given a temporary increase in base pay for the period of time they are active Black Belts.
- Black Belts are paid an additional amount when a project successfully completes one of the steps in DMAIC.
- Successful teams are given a fraction of the savings from working on a project. Distribution among team members can vary widely. Some organizations have even let the team decide how the money should be distributed.
- And many others—don't let our list discourage you from your own effective practices.

Our message here is that Black Belts typically command a premium in the marketplace over other, essentially identical project managers who are without the Black Belt competence. If you do not have a plan for Black Belt compensation you may find that every time you train a new Black Belt they get hired away by someone else.

SOURCES FOR OBTAINING BLACK BELTS

Train Black Belts

Thus far we have considered Black Belts as coming from within your organization. We think that this is the best option, but there are other approaches, discussed in the following sections.

Hiring Black Belts

When Black Belts are hired from outside the organization, they may not understand the organization as well as those who are part of it. This has benefits as well as drawbacks. They are not hindered by past organizational boundaries and prejudices. On the other hand, they may not be expert in the underlying technology of your business or understand some of the organizational boundaries, cultural expectations, procedures, and policies.

Using Outside Black Belts

There are a number of outside sources of people who are highly qualified in the area of Six Sigma. They will lack the organizational experience, and may or may not understand your specific technology. For this reason we recommend a team leader from within the organization (probably on less than a full-time basis), coached by the Black Belt. If your strategy is to use outsiders as Black Belts, many of the long-term benefits of Six Sigma will be lost. You will not develop the capability of Black Belts as an integral competence within your organization, and the future leaders will not have the experience of having led successful Six Sigma projects. All of these considerations lead us to characterize this approach as settling for half a loaf when the full loaf is within your grasp. At best we see this as a stopgap measure to get specific projects moving early, while you identify and train Black Belts from within your own organization.

Other Sources

- Use consultants as Black Belts. Most consulting firms in the area of Six Sigma will offer training as well as Black Belts to work on specific projects.
- Develop an alliance to share Black Belts. Some community or regional business development centers have arrangements for Black Belts. This can reduce your effort in finding a suitable consultant.
- Use Black Belts from a large supplier or customer. Some of the very large companies that are implementing Six Sigma will provide a Six Sigma Black Belt or Master Black Belt on a full-time basis to key suppliers or customers.
- Use a combination of these methods. Obviously you are not limited to any one of these. It is possible to use more than one. Coordination becomes an issue.

CONTINUING DEVELOPMENT

Any Black Belt who ends his or her development with the training provided in a Black Belt training class is not very motivated. Continued development and growth is expected as a normal course of events. There are a number of sources for the continuing development of Black Belts. A few of these are books, videos, seminars, software tutorials, specialist classes, universities, other Black Belts, and Web-based content.

In addition to the project deliverables for a Black Belt's project, one of the performance criteria is continued development and growth. As part of the budgeting and the enabling processes for a successful Six Sigma implementation, the continuing education, development, and growth of Black Belts must be considered. One of the questions is always where this expense is addressed. Here are some ideas:

- Costs are allocated against each project assigned to the Black Belt.
- An enabling account collects these expenses and allocates them across the business.
- The home departments of the Black Belts absorb the cost.

Black Belts must focus on fulfilling the needs of their Six Sigma partners, participants, and trainers, by becoming more productive, getting into and out of training quickly, and becoming more confidently responsive to the organization's needs. The training delivery teams help stakeholders to experience less painful training. For example, they might offer a wide variety of training delivery systems, depending upon what is most cost-effective and produces the "be able to do" competence desired by their Six Sigma alliance partners and stakeholders.

Training is hardly ever a solution by itself; rather, it is integrated into the organizational environment where expectations are clearly defined and measured, and its relationship with all other aspects of the work environment are understood and taken advantage of. This is important because, at those "moments of truth" when there is no one to ask and your employees need to take action, they will either help or hinder the achievement of your organization's desired business results.

We recommend process application with a specific project as part of the training and coaching cycle. This way the candidates are actively working on real projects as they are being trained. Nothing sharpens the ability to learn so much as having a need for immediate application.

We recommend a sequenced training process be used for developing Six Sigma Black Belt participant stakeholders. This approach uses twenty days, or 160 hours, of training. The training is normally performed in four one-week blocks, separated by several weeks of project activity. The expected time to implement this demonstration will be dependent on the project selected, the people involved, skill levels, and so forth. We recommend that the executive management team select this first project based on criteria that include high value to the strategic plan, but the project should be capable of completion by this new team participant within a six-month time frame.

Typically, project work selected is in the three- to nine-month time range by design. The list of modules found in Appendix C provides the Black Belt with the understanding of the philosophy of Six Sigma, how Six Sigma ties to the strategic plan, the use of Six Sigma as a metric, project management, and the process of DMAIC. In addition there is a balance of statistical and process learning with team leadership and understanding behavior.

Most new Black Belts need coaching by a Master Black Belt or instructor for at least the first project and most likely for the duration of the time spent as an active project-managing Black Belt. Everyone needs a coach. Not all of the tools and techniques are used on every project. When it comes time to use a tool or technique that has not been used before, it is good to have someone to offer expert advice. Coaching is a very important activity and requires dedication and effort. Chapter 8 is dedicated to coaching. Black Belt training program recommendations can be found in Appendix C, "Training Certifications."

10

GREEN BELTS

There are major classifications of people within the organization who need to be competent at Green Belt level. These are managers, Six Sigma project team members, and Green Belt project leaders. The speed for each of these classes to achieve Green Belt competence is dependent upon the individuals and the needs of a project (Figure 10.1).

Training for these groups can be done separately or in combined classes.

Managers

- Some organizations require all managers to be Green Belt qualified.
- Some organizations will not promote a manager unless he or she is Green Belt qualified.

Six Sigma Project Team Members

- To participate effectively on Six Sigma project teams, the members need to be Green Belt qualified.
- This training of team members is an important part of the timeline and budget considerations for a Six Sigma project.
- Green Belt team members provide the subject matter expertise for the project. Their experience in working in the process provides insight and perspective that cannot be gained in any other way.

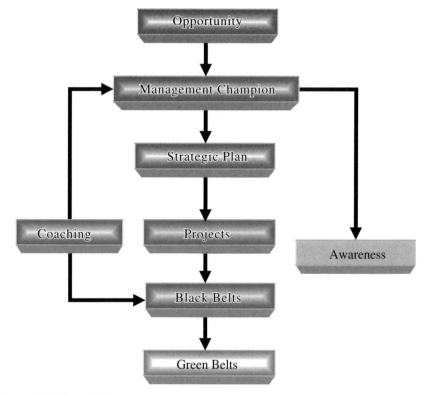

Figure 10.1 Green Belts.

- Most Green Belt team members are less than full-time on a project. They have a "normal" job to do and work on the Six Sigma project as a separate assignment. This requires that issues such as overtime, normal work demands, and so forth be addressed.
- Green Belt team members can come from any level in the organization; the key is that they use the process that is part of the Six Sigma project and have subject matter expertise and knowledge.
- Because Green Belt team members and their peers will have to live with the changes that are made, it is very important to have their input and support for any change.
- Green Belt team members can "sell" the project changes to their peers in a way that no management or technical leader can.
- Green Belts assist in benchmarking current Six Sigma project issues and concerns.
- Green Belts press goal achievement processes into play, creating road maps to success, providing recognition of arrival and "how to" maps for achieving the desired business results again.

Green Belt Project Leaders

For limited projects within a single department or function, a Green Belt may serve as the project leader. The fundamentals needed to lead a basic project are included in Green Belt training. Usually these are people who have successfully served on a project with a Black Belt project leader. The training for these people is the same as for all other Green Belts.

GREEN BELT TRAINING

Team members on Six Sigma projects receive their training immediately prior to working on a Six Sigma project team, or as a second alternative during the actual project. They have immediate application for the tools and techniques they are learning.

Green belt training is designed for people assigned to projects to work with Black Belts.

Approximately one month into the Black Belt training, Green Belt training efforts begin as needed for those who will work on the project teams on a less than full-time basis. We firmly recommend participants be assigned to a project before starting the training. Six Sigma Green Belt training applicants gain first-hand experience in selecting projects and using the fundamental tools involved in mapping and improving a Six Sigma process.

Participants will use their personal projects in learning how to follow the major steps of define, measure, analyze, improve and control (DMAIC).

Statistical techniques used are fundamentals only, and the use of a computer is NOT necessary for this training. The first week of the program consists of classroom training on tools and techniques of the Six Sigma processes. During the next three weeks, participants are required to apply these skills to personal or job projects. Participants then return for the final week of classroom presentations of remaining course materials.

When people are selected to serve as Green Belts on a Six Sigma project, a number of administrative and logistical questions may arise. We strongly recommend that you make decisions on these issues in advance. Some of the questions that have caused issues include:

- How many projects can be assigned to a department or function over a given time frame?

- What percent of a department or function workforce should be serving on teams at once?
- Can a Green Belt be on more than one project team at the same time?
- Does everyone eventually serve on a team and become Green Belt qualified?
- Are Green Belts appointed or volunteers?
- Who will do the work the individuals who are Green Belts normally do, while they are in training and serving on a project team, even on a part-time basis?
- If overtime is involved, where does it get charged?
- Will Green Belts receive extra compensation for a successful project?
- Will people who picked up the slack while a Green Belt participated on a project be compensated? If so, how, and how much?
- What is appropriate recognition for serving on a Six Sigma project as a Green Belt? If a range is possible, who makes the decision?

One caution: do not allow so many projects needing Green Belt team members that you exhaust your people, or they begin to neglect the "normal" work because of all of the project work.

Here are some training delivery options for Green Belt team members:

- Single-company training
- Training for each team, with Black Belts helping with instruction
- Multicompany training, with direct competitors in separate sessions
- Different locations to reduce travel
- Online self-paced training

Sources of Green Belt Team Members

We strongly recommend that you develop Green Belt team members from within your organization and not try to hire this level of competence from outside. Team members provide the knowledge of working in the process for a significant amount of time. They must ensure that the project does not get too far out of line with what is possible with the people in your organization. A caution, however: This role can be one that tries to prevent any change at all. Six Sigma projects will challenge the current way of thinking in your organization. The balance is to ensure that the changes are not so large that the organization rejects them out of hand. Things may have to be done in phases so people can make a change they see as possible, before moving to the next step.

Projects Led by Green Belts

When Six Sigma is well accepted in an organization and management has demonstrated its full support and commitment, it is natural for some smaller projects to be led by Green Belts. Criteria for Green Belt projects are less demanding than for Black Belt projects. Green Belt projects are usually contained within a single department or function and are not as formally scoped as the Black Belt projects. They can be a very effective way to share some of the Six Sigma practices and learning with people who have only had awareness training but are impacted by the changes made with Six Sigma. Use the same DMAIC steps for Green Belt projects as are used for the Black Belt projects.

Since the amount of training is less for a Green Belt than for a Black Belt, Sponsors will need to be more careful in the selection of the projects. Do not assign a Green Belt as a project leader when it is apparent that some of the necessary tools and techniques may not have been covered in their training. One way to handle this is to be sure to have a Black Belt, Master Black Belt, or other coach immediately available for the Green Belt project leader. In most cases where Green Belts lead projects, it is on less than a full-time basis. Some special allowances are made in work assignments, but the project is not a full-time assignment.

We are not particularly fond of having Green Belts lead projects in the early stages of Six Sigma deployments. If there is very much of this, the message is sent to the organization that you do not really need Black Belts, and being the project leader becomes a part-time job. Whenever people have two part-time jobs, one suffers. "No servant can serve two masters: for either he will hate the one, and love the other; or else he will hold to the one and despise the other." In our experience, without exception, it is the project that suffers. When there is a Green Belt project manager, we believe that the supervisor or manager must be the Sponsor of the project. This way, if there is a conflict between the two job functions, the Green Belt project manager can go to a single source for help in resolving the conflict.

Manager-Level Green Belts

Managers who need to have more than just an awareness of Six Sigma will be in Green Belt classes, but they may not be on a Black Belt project team. The objective for these classes is for managers to become conversant with the language, use some of the tools and techniques, be able to understand the Black Belt project reports, and have a better idea of the kind of team members needed for Six Sigma projects. It is best if they actually serve on a project team.

Often the projects for managers involve high-level teams for processes that cross many organizational boundaries.

Managers who are Green Belt qualified:

- Better understand Six Sigma process.
- Understand the language of Six Sigma from doing a project.
- Provide improved selection of team members for Black Belt projects.
- Use some of the tools and techniques on their own.
- Are better able to identify potential projects.
- Lead improvement efforts within their span of control.
- Learn to think about processes.
- Understand why defects are so expensive.
- May be become process owners or Sponsors.
- Learn team leadership and team-building skills.
- Understand the social style of individuals.
- Acquire meeting-management skills.
- Have the chance to work on communication skills—written and oral.
- Learn the goal achievement process.
- Acquire competence and confidence in using the various tools and techniques, which will advance their project management and daily management skills.

Training Delivery for Managers

An effective option for managers is to make the training available online and somewhat self-paced. Our recommendation is that there be some firm deadlines for completing prescribed modules of training. It is too easy to put things off for the more urgent daily business. Other options include public classes and having managers participate in training for Six Sigma project team members. Once the initial training is complete, however, an issue that must be considered is the training of the people who were not in the company during the original training, or did not get the training required for their current responsibilities.

If your organization makes Green Belt competence a requirement for promotion to manager level or for any promotion for existing managers, the training needs to be available on a regular basis. This is where the self-paced programs have some real advantages, especially for smaller organizations. If you have Green Belt competence as a requirement for managers, it is our recommendation that there be a clear communication of that policy, with adequate opportunity for people to avail themselves of the training. Once the cut-off date has been reached, there are no exceptions. We have seen this cause some real damage to the credibility of top management, when a vice president was promoted to senior vice president without Green Belt competence. In short, do not make Green Belt competence a requirement unless you are going to live by the policy for everyone, including yourself.

Green belts must focus on fulfilling the needs of their Six Sigma partners, participants, and trainers, by becoming more productive, getting into and out of training quickly, and becoming more confidently responsive to the organization's needs. The training delivery teams help stakeholders to experience less painful training. For example, they might offer a wide variety of training delivery systems, depending upon what is most cost-effective and produces the "be able to do" competence desired by their Six Sigma alliance partners and stakeholders.

We suggest considering the outline found in Appendix C as a model. This approach represents ten days, or 80 hours, of training, which is normally performed in two one-week blocks, separated by several weeks of project activity. Green Belt training is a subset of Black Belt training. The expected time to implement this demonstration will be dependent on the project selected, people involved, skill levels, and the like.

We recommend that the executive management team select this first project based on criteria that include high value to the strategic plan, but the project should be capable of completion by this Green Belt team participant within a six-month time frame. Typically, project work selected is in the three- to nine-month time range by design.

Green Belt competence will prepare individuals within the organization to use a variety of tools and techniques to define, measure, analyze, improve, and control. These skills are not limited to use on formal projects. As the use of the tools and techniques becomes more and more common, the quality of decisions within the organization will improve.

Green Belt training recommendations can be located in Appendix C, "Training Certifications."

11

AWARENESS

During Six Sigma deployments, plans need to be in place to provide every employee in the organization with a basic understanding of Six Sigma and how it will affect them (Figure 11.1). Why do we feel so strongly that organizations deploying Six Sigma initiatives should conduct awareness training for all employees, even those not assigned to a project team? Studies show that 68 percent of customers who quit using a business's products and services did so because of poor "employee attitudes." It seems logical that every organization should have multiple programs, with measurements, to ensure that "employee attitudes and behaviors" are as good as they can be. The employee attitudes in an organization can often be the one thing that dominates the success or failure of the business results. When employee attitudes are good, they become a foundation for appropriate employee behaviors, goal accomplishments, process improvement, and desired business results. Typically, a good attitude starts at the top of the organization and is reflected all the way down to the newest or lowest employee job function. It is difficult to build a successful organization on a poor foundation.

Good attitudes that are modeled and demonstrated at the top management levels are one of the critical parts of a good organizational foundation. When an employee's attitude is bad, it tends to conflict with and blot out everything else. Desired behaviors can often become obscured, goal achievement is likely to be forgotten, process improvement is ignored, and good results are difficult to find. Can you personally list your organization's "Employee Attitude and Behavior" strategies? Employees not trained as Black Belts or Green Belts may serve

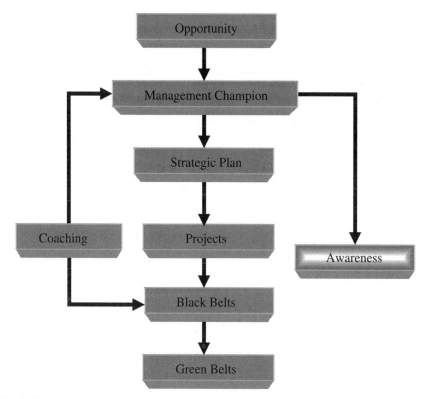

Figure 11.1 Awareness.

on other teams and thus are required to complete the organization's specific Six Sigma awareness training program. The organization may have teams known by a variety of names, efforts, and levels of autonomy, such as Natural Work Teams, Quality Circles, Kaizen Teams, and Continuous Improvement Teams.

Whatever the team name, adding awareness training is a good way to gain returns on your Six Sigma deployment investment. Instead of just telling employees about your Six Sigma deployment process and adopting a "sink or swim" approach, you might consider a more effective way. It takes time, effort, commitment, and adherence to the values by all concerned, but the long-term results are worth it!

One learning theory teaches that if you meet with failure while attempting to accomplish a task, you must make adjustments and try again. This process is repeated until you are successful. A behavioral model, which we call the

Learning Cycle, includes the following sequential steps:

1. Tell individuals specifically how to perform certain tasks.
2. Show individuals specifically how to perform these tasks.
3. Allow the individuals an opportunity to repeat the desired performance.
4. Reward the performer/trainee for accomplishing the desired performance.
5. Redirect individual performances that are at a less than appropriate or desired level.
6. Repeat this "cycle" until the individual trainee can perform the desired task 100 percent correctly.

Six Sigma awareness training makes a major contribution to an organization in several fundamental ways. First, Six Sigma awareness training focuses the organization's attention on the customer and the issues that are critical to the customer. Paying attention to the Customer's Critical Criteria results in products and services that meet the customer's desires. When the organization stakeholders meet the Customer's Critical Criteria, market share is maintained or expanded.

The second major contribution is internal to the organization. Six Sigma awareness training identifies why defects must be driven to a very low level. As defects are reduced, productivity is increased, costs are reduced, issues receive a faster response, and additional capacity is often discovered. Once the "hidden factory" is put to productive use, it can often reduce capital spending while increasing capacity. Six Sigma goal achievement and process improvement become integral parts of all organizational activities directed at achieving the desired business results.

Six Sigma awareness training assists in the skill development of all organizational stakeholders, regarding the proper use of problem-solving tools and techniques. Awareness training often assists in the development and use of cross-functional teams that provide employees with a better understanding of the organization and its strategic plans. Six Sigma deployment awareness training enhances attitudes, awareness, knowledge, and essential skill levels of the organization's work force. People must be capable of using the tools and techniques provided. In some cases this is a matter of training, and in other functions or areas of responsibility a general level of education may be required.

Six Sigma awareness training makes it clear that systems are the way processes are linked together in order to accomplish a desired intent. Participants in awareness training discover that procedures are the specific step-by-step activities that are performed in order to achieve the desired results. Awareness training teaches participants that:

- Projects are to be evaluated on their contribution merit to an organization's strategic plans.
- Tools and techniques are necessary, but will not deliver without the people behind them.
- Those bad systems too often defeat good people.
- People development is more than teaching tools and techniques.
- Synergy between strategy, people, and systems with appropriate tools and techniques yields more than the sum of its individual parts.
- Once your customers have developed a low tolerance for defects, you have a unique marketing advantage over your competitors, provided you continue to focus on the Customer's Critical Criteria and improvement processes.

Many businesses would like to improve their people, processes, and systems to attain greater market share and profitability. Some business are behind the technology curve, but still do not invest in training their employees, offering the excuse that they can ill afford the loss in productivity and high cost factors often associated with traditional training processes. Yet others fail to recognize that providing Six Sigma awareness training to their employees often means building a competitive business advantage that results in satisfied customers, fewer mistakes, less rework, and greater profits to the bottom line.

We believe in providing Six Sigma awareness training to teach your employees that excellence, innovation, and anticipation are essential to an organization's ability to achieve world-class performance. Awareness training has a focus on your customers. Make this "real" and back it up with training skills for using the attitudes and behaviors that place the customer's needs first. Tie your organization's business systems and strategies to your customers' expectations and desires. Awareness training points out that a data- and fact-driven management will provide effective measurement systems to track both results and outcomes, as well as process input and related predictive and preventive data.

Awareness training demonstrates that an executive management team that focuses on the organization's process improvement will be the driving force for the growth and success of the organization's desired business results. Improvement processes will be documented, communicated, measured, and refined on an ongoing basis. Awareness training illustrates that proactive management involves habits and practices that anticipate problems and changes, apply facts and data, and question assumptions about the organization's goals and "how you do things." Awareness training must explain that the organization's "no boundaries" collaboration requires cooperation between all internal customers, external customers, suppliers, and other stakeholder partners. World-class performance and perfection require a tolerance for failures, giving internal customers (stakeholders in the organization) the freedom to test new approaches, even while managing risks and learning from mistakes. Six Sigma awareness training that successfully embraces these essential improvement themes will raise the performance and customer satisfaction bars for your organization.

Gaining continuous and incremental improvements will put a firm into a better position to successfully achieve its desired business results of gaining and maintaining loyal customers, improving market share, and increasing bottom-line profitability. The efforts to learn and apply Six Sigma may not be necessarily easy or painless. Participants of Six Sigma awareness training understand that not every tool, technique, or approach needs to be used on every project. Part of the success of Six Sigma is using only those tools that are needed for a specific project. However, during the training we encourage you to apply as many Six Sigma tools and techniques as possible.

Participants of awareness training understand that the organization's strategic business plans are grounded in recent experiences and clear thought, that they are concise, actionable, and measurable, and that they contain the following elements:

1. Business review. The first step in the business plan is to review your firm's performance over the last year, identifying the sales and profit status, essential accomplishments, and essential opportunities for improvement.
2. Current year objectives. Business objectives are focused desired results in the areas of sales and profit. Objectives consider historical performances. All objectives are written, so that it is easy to measure progress over time.
3. Core strategies. Determine the core strategies that will guide the business in the current year. Strategies include "What to Do," and tactics include "How to Do It."

Participants of awareness training understand that Six Sigma is an aggressive performance goal for almost every process, but that there is a point at which you move on to other, more productive efforts. Six Sigma places a premium on application. Significant improvement in business results is expected from each project. Six Sigma anticipates significant improvement (change). Since projects are well scoped and defined to align with the strategic objectives of the organization, the changes are necessary and important. Successful Six Sigma implementation plans do not create a separate bureaucratic organization. Keep the Six Sigma people in the business units, with the business unit having accountability for achieving results.

Participants of awareness training understand that Six Sigma can be applied to any part of the organization. In fact, organizations that have applied Six Sigma to the manufacturing division find that the largest gains are frequently outside of manufacturing. Remember that the customer is king. The Customer's Critical Criteria are required, and they enhance customer loyalty, yet too few know what these criteria are for each customer.

Participants of awareness training understand that the "process" aspect of Six Sigma methodologies is fundamental to any business process. Most of the process steps include common sense and are arranged in an order that has been proven to make them effective. Businesses that are not practicing Six Sigma are likely to experience huge disconnects between their customers' expectations and the delivery of their products or services.

Awareness training considers the strengths of these Six Sigma themes:

- Focusing on your customers
- Data- and fact-driven management
- Process management and improvement focus
- Proactive management
- Collaboration with all stakeholders and without boundaries
- Striving for perfection, with a tolerance for failure as a learning process

Participants of awareness training understand that cost control is a major concern for most companies. Why does a multifaceted objective such as cost control and reduction involve every aspect of the organization and every employee? If only top leadership is concerned about cost control, they will spend a large portion of time trying to police the activities of others. Management often attempts to place controls on everything imaginable, just to be sure that the costs do not get out of stated alignment.

Participants of awareness training understand why costs frequently continue to increase.

Often we have an ongoing emotional battle because people both inside and outside our organization seem to be conspiring to run up our costs. To accomplish cost control objectives over a long period, we apply the interaction of three essential areas: strategy, people, and systems. It is the interaction of these areas that allows the employees (internal customers) to respond appropriately in those moments of truth when they face decisions that either decrease or increase your organization's costs.

Participants of awareness training understand that we must collect all needed data and transform it into useful information before we can understand our cost issues and make needed changes to them. Setting up the measurement systems and understanding what the data are telling us become critical to any kind of cost control effort. If we wait until the financial numbers arrive, usually it is too late for us to be able to use the data as an effective cost control methodology. Measurements that are closer in real time are required, if the data are to be of any immediate use. It is easy for a situation to get out of control before the end of the financial reporting cycle. Knowing the accuracy and precision of the data becomes a concern as these real-time measurements are established. Efforts to implement lasting cost control measures seldom maintain their effectiveness when done in isolation. These measures need to be part of an overall organizational focus.

Participants of awareness training understand that when there is synergy of the three elements of strategy, people and systems, it is more likely that cost control efforts will become a part of the basic culture of your organization. Once embedded in the culture, efforts to make cost control an ongoing part of the organization are easily enforced as measurements are tracked. Internal customers will understand what is important in terms of goals and where the organization is headed, if they have a well-developed strategy that is communicated and "lived" by the leadership of the organization. With appropriate people development, the attitudes and behaviors of the "worker bees" are tuned to the strategy of the organization.

Participants of awareness training understand that the "worker bees" who do not agree with the strategy are likely to self-select to leave the organization. This is a benefit to both the individual and the organization. Only those who understand and support the strategy will make the kinds of decisions that are appropriate for long-term business results. Focused strategy and supportive people require systems that allow them to do what needs to be done in the most cost-effective manner. Whatever the strategy, if six signatures and a requisition in triplicate are required to purchase a pencil, then there is little chance that employees will be an active part of any cost control effort.

Participants of awareness training understand that when the "strategy, people, and systems" action plans are working in harmony through the internal customers of the organization, it is difficult not to become more cost-effective. Your cost control efforts should then be focused on continued efforts to ensure that the strategy is correct, that people are developed, and that systems are performing as needed. There are numerous ways to establish a competitive advantage; the key is that the customer is the one who decides if what you consider a competitive advantage really is an advantage. When Coke decided that a new taste would offer them a competitive advantage, they introduced New Coke and discontinued the old recipe. Only after severe customer complaints and loss of market did they return with Classic Coke. No one can be all things at all times. Those who try generally become third rate at almost everything. What is it that provides your organization with a competitive advantage? Spend some time confirming this with your employees and, more importantly, with some of your customers.

As you start preparing for Six Sigma awareness training, realize that there are no magic bullets. This is work and effort applied in a planned and intelligent fashion. When projects are completed, one at a time, there is an additive effect. Improvements will tend to be linear with each project that is completed. Exponential growth is possible, and the project pace can increase as organizational stakeholders gain more project experience and success. A truly successful Six Sigma project effort shares the success and challenges. This not only gives encouragement to others, it also allows people with success stories to receive some of the recognition they deserve. Every project is a learning experience.

Awareness training helps start the focus on the customer, and it is likely that the processes involved will cross the functional or divisional boundaries in your organization. Processes that cross departmental boundaries frequently have the biggest issues at the boundaries. Waste, scrap, rework, redundancy, inattention, shifting the blame, and so on are likely to be found there. Within the departments or functions, internal bureaucracy frequently rules. Suboptimization to meet the department goals without consideration for the impact on the ultimate customer or the overall organization is common. It is this crossing of normal boundaries that will present Black Belts and their teams with some of the largest challenges. Champions and senior management need to be constantly aware and supportive of improvement activities that are multidepartmental or multifunctional in nature. These projects also require the greatest amount of interpersonal skill and understanding on the part of the Black Belt.

The Six Sigma awareness training process starts to remove the fear and mystery of using statistics in the improvement process by exploring definitions, presenting interesting facts, and providing usable skills training. Statistics are a vital part of our everyday lives. Most of us use statistics to make important decisions about our everyday lives. We are bombarded by news of statistical reports regarding the weather, society in general, health, insurance, mortality, birth rates, economic conditions, growth in our communities, and so on. Can you think of some recent statistical reports to which you have been exposed?

Participants of awareness training understand that statistics are collections of quantitative data. A statistic is a single term or datum in a collection of statistics. A statistic is a quantity that is computed from a sample, such as a mean, average, or sigma. A statistic is a random variable (one that has variation or is subject to change) that takes on possible value as information. The term "statistical" means "related to or employing the principles of statistics." Thus, we use statistics in our daily lives to make more informed decisions. Why not do the same to improve our desired business results?

Many of us, over the course of the last few decades, have been involved in various types of statistical improvements to make our organizations more competitive, to claim larger market share, and certainly to improve profits. Many organizations that have used such programs have experienced some impressive and positive results. However, for the most part, organizations found that the "big ticket" items of their desired business results did not mature. Why? Was this the fault of the program's limitations, lack of stakeholder skills, or maybe management's lack of total commitment? Maybe it was a combination of all of these. What makes Six Sigma so different?

Participants of awareness training understand that Six Sigma melds the business organization's focus through strategy, binds the synergy of people and process, then aims these at world-class performance in the areas that are critical to the organization's customers. Utilization of statistics is certainly a large and important part of our Six Sigma comprehensive improvement processes, but it is only a part. Six Sigma is more than simple process improvement. While the tools and techniques of process improvement are vitally important, if these tools are not supported by an underlying business strategy and the development of people, the gains will not be as big as they could be.

Participants of awareness training understand that Six Sigma focuses on the organization's Customer's Critical Criteria and world-class performance, with the process improvements, people development, and sound business strategy necessary to yield results. Six Sigma will provide you with the tools and techniques to support your organization's business strategy. It will show you how to measure, analyze, improve, and control processes. You will gain knowledge of people dynamics, including a better understanding of yourself, which is essential for process improvements to succeed.

Participants of awareness training understand that too many times there are attempts to focus on process improvement, or people development, using one to the exclusion of the other. Success with these approaches is usually transitory. Long-term sustained improvement requires both. Six Sigma priorities begin with a focused understanding of your Customer's Critical Criteria, expectations, satisfactions, and values as they regard your organization. Performance is then measured against these customer criteria. Customers are internal as well as external to the organization.

Participants of awareness training understand that Six Sigma requires management to be data- and fact-driven. Management must clarify the measurements that are key to determining the success of the organization's desired business results and performances. Six Sigma is a process-focused management and improvement system. The process helps organizations to build competitive advantages and to deliver value to their customers. Six Sigma requires the organization's management teams to become proactive rather than reactive. Management establishes clear goals and a focus on problem prevention instead of fire-fighting emergencies. Management will be asking, "Why do we do what we do?" instead of just going along because that's the way it's been done in the past. Six Sigma internal customers benefit greatly by gaining knowledge as to how they might fit into the organization's big picture. Internal customer stakeholders are able to recognize and measure the interdependence of activities in all parts of a process.

Participants of awareness training understand that by gaining an understanding of the end users' needs and of the way how work flows through an organization's process or supply chain, the internal customers are better equipped to meet the needs and Customer's Critical Criteria of external customers, thus preventing costly mistakes, failures, or waste. Awareness training participants understand that Six Sigma projects are designed to improve, to change, or to successfully achieve the organization's desired business results.

Awareness training helps us discover that introducing change means to challenge the "old guard" of both personal and organizational paradigms. Paradigms are our personal beliefs about how things are supposed to be. We often filter out and resist incoming data that do not closely match our specific paradigms. As people and as organizations, we get set in our ways and become comfortable, and thus most suggested or required change makes us uncomfortable.

As a result, some of us will resist making some or all of the needed changes. Once we have personal buy-in to the new change, we can begin the slow process of change. We say slow because those wanting the change to take place generally feel that the rest of us are not moving towards the change fast enough. Certainly each of us has been asked to change and have asked others to change as well. How did you feel about your progress or their progress toward the desired target? The greater the belief in our paradigms, the more difficult will be the opportunity for making needed change. Much resistance from personnel within the organization is often the result, as change "levels the playing field" and some stakeholders feel they will lose their personal or positional power.

Participants of awareness training understand that most products or services that do not change and evolve are likely to have limited success and life in the marketplace. The auto industry, for example, has locked itself into a cycle of a new model every year. Frequently there are only minor or cosmetic changes made to these new models. Ever have trouble identifying one year's model of a particular vehicle from the next?

Six Sigma awareness training teaches that the change or redesign of our organizational products and services is based on the measured value according to our customers' needs and our suppliers' capabilities. Performed correctly, the changes made to our products and/or services do a better job of meeting and exceeding our customers' expectations than did our original product or service.

When Six Sigma deployment is implemented by an organization, the economic return on investment will improve for the business units involved. Better products with fewer defects, cost reduction projects, better understanding of the customer, improved workflows, employee commitment, supplier involvement, and reduced cycle times are but a few of the contributing areas. Good redesign or change processes reduce complexity, involve fewer defects, improve reliability, and offer the customer functionality not found in the original product or service. When these objectives are met, they have a direct cause and effect relationship with improved margins and larger market share.

Awareness training explores why implementing Six Sigma requires taking a new and closer look at many of the essential parts of your organization. Six Sigma implementation must have some structure within your organization; it cannot be a bolted-on activity that is managed outside of the normal business. The business units of your organization develop Champion and Black Belt personnel to ensure the successful achievement of your organization's desired business results. Senior management of the business units must maintain responsibility for the success of Six Sigma. Champions must have ready access to the senior management and be knowledgeable of the activities of the Black Belts in their business. Six Sigma projects are not started until the Black Belt and Champion agree that the scope and parameters of the project are adequately defined. It is then up to the Black Belt to lead the project to its completion.

Participants of awareness training understand that people development is more than teaching tools and techniques. Leadership and team-building skills are essential for the Black Belts. Understanding behaviors and providing motivation are equally important development areas. Each component of your business system is an integrating factor for each other component within the system. First evaluate your business system as a whole, and then move to exploration of the components. The links between components are as important as the components themselves. Complexity can overwhelm and undermine your efforts. Seeing the patterns behind the events can contribute to simplification and understanding. This, in turn, can give you a much-needed advantage in your actions aimed at changes to improve your organization. It also helps you to identify limits to your success, or growth problems within your organization.

Participants of awareness training understand that without complete, appropriate information and analysis of the entire system, you set yourself up for failure. Explore "systems" thinking, viewing your business world as a whole, a framework of patterns and interrelationships, in order to successfully achieve your organization's desired business results. If the only indicators in the Six Sigma organizations are financial, they may be pushing a ball uphill, and with a time bomb ticking. If these Six Sigma organizations divert attention to anything else, the ball rolls back down and they must go get it and push it back uphill again. This is a constant fight against gravity. Instead of having to fight to lift every accomplishment, with gravity making it more difficult, develop processes and systems that make gravity work for you. If the process is the best way for all concerned, including the organization as a whole, then once the process is started, you only have to get out of the way and allow it to work. If a process demands continued unusual effort and attention, it probably has not been developed in an optimal fashion.

We feel that we have provided you with some material and guidance regarding Six Sigma awareness training. However, nothing creates the impact as much as the leadership of the organization taking the time to ensure that all employees have a basic level of understanding about Six Sigma. This is an excellent opportunity to share the strategic plan, vision, mission, objectives, essential performance indicators, goals, and plans.

Depending upon the amount of effort you wish to make and the level of training you require for employees, awareness training ranges in time from half a day to two days. If you can afford the two-day period, you will provide employees with some explanation of the fundamental improvement tools. Awareness training is important for every member of the organization for several reasons. Among the more important are that it allows you to:

- Satisfy people's desire to be aware of things and know what is going to happen.
- Convey that world-class performance, as defined by the customer, is what we are seeking to achieve.
- Ask for cooperation and help from the employee population.
- Start to generate ideas for potential projects.
- Prepare for change.
- Introduce your Customer's Critical Criteria as an important consideration.
- Explain how any failure to meet the customer criteria becomes a defect.
- Effectively share the strategic plan, vision, mission, objectives, key performance indicators, goals, and plans.
- Introduce the language, concepts, methodologies, and some tools associated with Six Sigma processes.
- Introduce the process for project selection.
- Demonstrate leadership participation and commitment.
- Solicit volunteers for Black Belt and Green Belt.
- Start to break down barriers between departments, functions, and levels in the organization.
- Encourage systems and process thinking.
- Start to give employees an appreciation for the impact of variation.
- Answer the question What Is in It for Me? (WIIFM).
- Introduce the essential players for Six Sigma implementation to your organization.
- Explain how each individual can sharpen his or her customer focus.
- Encourage systems and process thinking.

When the awareness training is completed, employees understand how to improve the business systems and processes using process improvement tools, and how to develop the capability of people as individuals, in teams, and as an organization. They are aware that efforts to improve either process implementation or people development at the expense of the other will not be as effective as will an approach that considers and includes both.

These employees know that the result of successful Six Sigma application is the melding of the business organization's focus through strategy, which binds the synergy of people and process, then aims these at world-class performance in the areas that are critical to the organization's customers. They are familiar with DMAIC (define, measure, analyze, improve, and control), the major steps for a Six Sigma project. Participants understand that within each step there are specific deliverables due before one can move to the next step in the process. Participants also understand that specific tools and techniques are especially useful for each specific step (Figure 11.2).

Employees with awareness training know how tools are used in Six Sigma:

- The goal is to use the tools that are necessary to accomplish the tasks.
- Use of the tools is not the desired result; the tools are learned in order to make the process more efficient and effective.
- Do not use a tool if it is not needed for a specific project. Make the tools work for you; don't work for the tools.

Because participants in awareness training know about the basic steps involved in a Six Sigma project and the tools used, they will better understand and

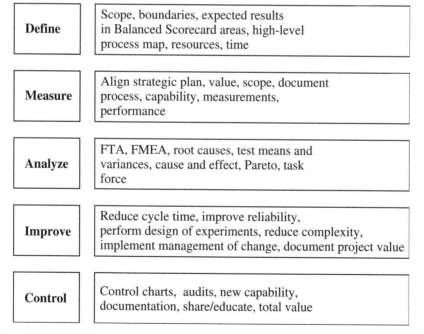

Define	Scope, boundaries, expected results in Balanced Scorecard areas, high-level process map, resources, time
Measure	Align strategic plan, value, scope, document process, capability, measurements, performance
Analyze	FTA, FMEA, root causes, test means and variances, cause and effect, Pareto, task force
Improve	Reduce cycle time, improve reliability, perform design of experiments, reduce complexity, implement management of change, document project value
Control	Control charts, audits, new capability, documentation, share/educate, total value

Figure 11.2 DMAIC.

be more likely to accept the changes that result. Some individuals are likely to start a self-education process to become more knowledgeable about the Six Sigma processes and the tools.

Awareness training will give employees the answers to questions such as:

- How many projects should there be in a department or function over a given period?
- What percentage of a department or function workforce serves on a team?
- Can a Green Belt be on more than one project team at the same time?
- Does everyone eventually serve on a team and become Green Belt qualified?
- Are Green Belts appointed, or do they volunteer?

While hard rules and procedure may not be necessary to cover all possible occurrences, some thought into general policies and practices will save problems later, if properly communicated.

As the people component grows, so will the competence and confidence in using the various tools and techniques. As the organization as a whole grows in understanding and applying Six Sigma, the size and complexity of projects can increase. Another advantage of awareness training is that the time needed for projects at a similar level of difficulty can be estimated faster. Green Belts can learn to do more of the project work, freeing Black Belts to lead additional projects. New employees need to understand the logic, reason, language, and tools of Six Sigma, and awareness training will provide them with that understanding.

Any successful project returns dollars in profit. We are in business for a variety of reasons, but without adequate profits, the business cannot sustain itself for very long. Six Sigma projects return significant dollars to the profit of the organization. These dollars can be clearly traced, using good accounting principles, from the project directly to the profits of the organization. Be sure to recognize all those who participate in a successful project. The rewards may vary—just remember that recognition and praise are infinity divisible. Be lavish with them!

Awareness training will not make an expert out of the average employee. It can give employees an appreciation of what is involved and provide them with enough understanding to recognize opportunities. With encouragement, some employees will start to generate ideas for potential projects.

Management should have the processes defined in advance to handle these suggestions and to explain how ideas will be evaluated. Key to any suggestion or recommendation from an employee is recognition and appreciation for the effort put forward to generate the idea. We recommend that this be done within two working days. Additional communications are in order as the idea is evaluated and acted upon, or added to the inventory for future consideration.

In some cases, there may be a strong interest to learn even more about Six Sigma, which will generate volunteers to serve as Black Belts or Green Belts. Be prepared for the range of possibilities, and define the process for selection of Black Belts and project team member Green Belts. While not all volunteers may immediately get on a project, there should be a note of appreciation for their willingness to serve. While consultants and outside trainers can help with awareness training, we consider it imperative that management take an active part in the awareness training efforts. Management may not do all of the awareness training, but at least one member should start every session. Topics for management to address include:

- Why we are deploying Six Sigma
- The expected benefits from Six Sigma deployments
- How projects will be selected
- How project team members will be selected
- How to turn in good ideas
- How to volunteer for a project
- What employees can do to help if they are not on a project

Awareness does not end with a single training session. Rather a planned series of communications reinforces the importance of Six Sigma to the organization:

- Announce to the organization when the first Black Belt training will start.
- Highlight when the first Black Belt training is complete.
- Provide an overview of each successful project. Include value to the organization, customer impact, and a brief review of the tools and techniques used on the project.
- Add up the value of Six Sigma projects across the organization quarterly.
- Send press releases to appropriate media.
- If your organization does not have a library, start a resource center that has a list of articles, books, summaries of projects, and the like, all related to Six Sigma. Make these available to all employees.

- Have Black Belts make periodic presentations of the successful projects they have led.
- Make use of all the signs, posters, pens, coffee cups, lapel pins, and the like associated with any significant effort in your organization.
- Attend and/or sponsor Six Sigma conferences.
- Provide a review of Six Sigma contributions in the Annual Report.

Employees are very discerning; they understand what is important to the management and leadership of the organization. Our behaviors give us away. There is a lot of truth in the trite saying: "What you do speaks so loud, I cannot hear what you say."

12

SIX SIGMA SUCCESS STORIES

MOTOROLA

Motorola's Six Sigma journey, which began in 1987, was started with defining the company's core values as vision, beliefs, goals, and initiative. The following are examples of some of Motorola's core values:

> Vision—Total customer satisfaction
> Beliefs—Mutual respect and uncompromising integrity and trust
> Goals—To be the best in class in every aspect of business that includes product development, manufacturing, marketing, and sales
> Key initiatives—Six Sigma quality and reduced cycle time

Having defined the core values, constancy of purpose was created through change management training of all managers for about two weeks, followed by mandatory training in the core programs of Understanding Six Sigma, Design for Manufacturability, Cycle Time Management, and Process Quality Improvement.

The managers then propagated the message and provided training to their employees. For example, product managers trained their employees in Design for Manufacturability. This helped in reinforcing the discipline for achieving excellence in various aspects of the business.

Success stories are based on authors' personal experiences and information gathered from websites for Motorola, General Electric, Honeywell, and Camp, Inc.

To benefit from the Six Sigma initiative, the most important of management tools Motorola employed was the establishment of a consistent, enterprise-wide, relevant, and accountable measurement system. Once the measurement system was in place, an aggressive goal-setting process was created. Accordingly, each quarter, employees would set some "reach out" goals beyond their normal responsibilities. Each group of employees, division management, and sector leadership would then review their performance on a weekly basis, with high expectation of achieving the goals. When extraordinary results were achieved, the team received a CEO Award for quality improvement. The award was presented by the CEO, Mr. Robert Galvin.

In addition to the CEO Award, savings that resulted from the exceptional work of employees were shared with employees, according to a pre-established and well-communicated formula. There was a clear understanding that Motorola, indirectly, meant Six Sigma. The world expected Motorola to be a Six Sigma company, a kind of Supercompany, with the best in quality on the planet.

In addition to the training provided by the managers to employees, Motorola University was staffed with employees and consultants to provide training to all employees when needed to support the implementation.

After initial successes internally, it was soon realized that Motorola could not achieve its ambitious Six Sigma results without partnership with its suppliers. Therefore, a supplier involvement program was developed to partner with suppliers at the early stage of product development. The supplier selection process was updated to promote implementation of Six Sigma at suppliers' facilities. Suppliers were even required to take the four core courses listed above. Suppliers tremendously benefited from the education and from partnership with Motorola. An expectation was created that if a supplier, irrespective of its size, was going to work with Motorola, it must understand Motorola's Six Sigma language.

With all these advances in management philosophy, methodologies, and leadership, Motorola saw a series of innovative products, the most successful one being the personal phone that revolutionized the telecommunications industry. The personal phone known as MicroTac set the stage for today's cell phones—small in size, light in weight, available in various colors and designs, easy to use, and ever changing.

Some of the initial success stories appeared in nonmanufacturing areas such as corporate accounting, printing, and customer service. In the manufacturing areas, the methodology was applied throughout the corporation, that is, at various divisions worldwide. Several success stories were published internally to publicize the success of Six Sigma.

The journey continued until 1992, the well-publicized date set by Motorola to achieve Six Sigma corporate-wide. Extraordinary results were achieved and Motorola grew strongly. Its sales almost doubled in five years, and this took place with approximately the same number of employees. Profitability improved from millions of dollars to hundreds of millions of dollars.

The benefits that were realized by Motorola have been documented in its training programs, as follows:

- Savings in manufacturing areas during the initial years were greater than $2 billion.
- Similar gains were expected in the nonmanufacturing areas.
- Response time improved dramatically.
- Profitability multiplied several times.
- Sales per employee almost doubled.

As the company was growing rapidly, its employee base started to change. New employees joined the company at a rate faster than its ability to communicate its commitment to Six Sigma. Besides, new managers from other industries were not tuned in to Motorola's commitment to quality and Six Sigma. As a result, the effort started to be diluted. Motorola has set a goal to reduce defects every two years, that is, ten times the improvement every two years, through the Six Sigma methodology.

GENERAL ELECTRIC

As Motorola's emphasis on Six Sigma was yielding results, GE committed to implement Six Sigma corporation-wide. Then CEO Jack Welch's passion for Six Sigma is well known. He embraced Six Sigma as a strategy to drive business process improvements and achieve improved profitability. GE's success with Six Sigma has exceeded most optimistic predictions. GE associates embraced Six Sigma and applied it to everything, including NBC operations. GE's Six Sigma built on its successes by sharing best practices across all of its businesses for better and faster customer solutions.

GE adopted Mikel Harry's BreakThrough strategy. GE had customized Motorola's Six Sigma training programs to develop Six Sigma Black Belts. Like Motorola, GE established its values as "Respecting Always the Three

Traditions of GE . . . Unyielding Integrity, Commitment to Performance, and Thirst for Change."

In addition to the above values, GE has identified passion for customers, meritocracy for people around the world, growth and globalization, innovation, speed, and excellence as initiatives, as a part of its corporate value system. Passion for customers requires measuring customer success that is driven by Six Sigma quality. To achieve their corporate objectives, GE has identified hallmarks of leadership that include Passion for learning and sharing ideas, and Commitment to delivering results in every environment.

According to GE, Six Sigma is a discipline of defining, analyzing, improving, and controlling the quality in everything at GE. The key concepts of GE's Six Sigma are the following:

- Critical to Quality attributes: those that are most important to the customer
- Defects: failing to deliver what the customer wants
- Process Capability: what we can deliver
- Variation: what the customer sees and feels
- Stable Operations: those that ensure dependability
- Design for Six Sigma: innovation to meet customer needs and process capability

GE's Six Sigma philosophy that customers feel the variance, not the mean, is an excellent explanation of quality improvement effort. Customers judge the perception of quality—not what the product does, but what it does not do. The variance results in defects and customer dissatisfaction. Six Sigma focuses on reducing process variation.

GE reported Six Sigma as a contributor to its growth, higher profit margins, and overall value to shareholders. The savings due to Six Sigma have been cited in hundreds of millions of dollars. Six Sigma has become an integral part of the GE Operating System, which consists of Globalization, Six Sigma Quality, Product Services, and e-Business.

HONEYWELL

Following GE's success with Six Sigma, Allied Signal and Honeywell independently implemented Six Sigma methodology. After their merger, the Six Sigma at the two companies became Six Sigma Plus, a proprietary system of the new company, Honeywell. Six Sigma Plus is considered a powerful quality strategy to achieve excellence. Six Sigma Plus combines the best practices of both companies, adds capability, and takes its continuous process improvement methods to a higher level of excellence.

Six Sigma Plus consists of the following tools:

- Voice of the Customer
- Lean Enterprise
- Enterprise Resource Planning
- The Honeywell Quality Value assessment process
- New skills and techniques for Total Productive Maintenance
- Broader applications for Activity Based Management

Honeywell has used Six Sigma Plus to empower its employee quality improvement teams to create more value for its customers and to improve its processes, products, and services. Besides being used in aerospace, polymers, chemicals, automation, control, transportation, and power systems, Six Sigma has been deployed in nonmanufacturing functions as well.

Honeywell has invested significantly to accomplish dramatic business results. All managers, supervisors, and professionals are required to become certified Six Sigma Plus Green Belts. Certifications for those with higher Six Sigma Plus skills also are part of Honeywell's Six Sigma Plus system.

At its Web site (www.honeywell.com), Honeywell has listed several examples of the success of its Six Sigma Plus methodology, as follows:

- A Six Sigma Plus team created the first and largest Internet auction site for used truck and automotive parts, generating more than $100 million in additional high-margin revenue for Honeywell within five years.
- A Six Sigma Plus team used the Internet to expedite customer payments and improved the cash flow by $6 million over fourteen months.
- A Six Sigma Plus team developed a metal injection molding process that enabled the production of a new, variable-weight golf putter, capturing a $10 million business for a Honeywell customer and $1 million in sales for Honeywell.
- A process improvement effort using Six Sigma Plus at a Honeywell Turbocharging Systems plant led to an increase in revenue by 300 percent.

Overall, Honeywell has reported about $2.0 billion in cumulative savings from Six-Sigma-related activities. Honeywell expects to save about $500 million per year using the Six Sigma Plus methodology.

Having experienced the success of Six Sigma Plus internally, Honeywell now offers its expertise to customers and suppliers. Honeywell makes available its Six Sigma Plus experts and its proven methodology, training, and consulting when needed. Six Sigma Plus in itself has become a revenue-generating service to its suppliers and customers.

One similarity between Honeywell's and GE's Six Sigma approaches is that both focus on implementing high-impact projects that drive results consistent with the needs and priorities of the business. Six Sigma Plus utilizes a project selection process that is linked to the company's strategic planning process.

Honeywell sustains the Six Sigma Plus culture at its facilities by investing in the training of its employees. Six Sigma Plus has fueled customer-oriented process improvements and innovative products and services throughout Honeywell.

For Honeywell, "Six Sigma is a vision we strive toward and a philosophy that is part of our business culture."

SUCCESSES AT SMALLER COMPANIES

Successful implementation of Six Sigma at smaller companies is less documented. Also, being privately owned, smaller companies are less willing to share financial results. The following section outlines a summary of how Six Sigma often evolves at a smaller organization.

The company CEO or owner learns about the Six Sigma methodology and its benefits at one of its successful clients. To some executives, Six Sigma appears to be a technique to reduce variation; to others, it is a strategic tool to improve financial results. To the rest, Six Sigma could appear to be another quality program. They've already had many of those by now.

Once an owner/CEO is convinced of the advantages conferred by Six Sigma and visualizes the benefits, it is much easier to implement Six Sigma at smaller companies and to realize its benefits. They send some of their managers to learn about Six Sigma, either by becoming certified Black Belts or Green Belts. The certified Black Belts report back to the management, fueling the Six Sigma fire. The CEO now wants to train the entire management team and wants to get started. An executive overview of Six Sigma is held, to develop a common understanding of the Six Sigma methodology.

At a smaller, but leading, manufacturer, it took more than a year to have its people trained, management oriented, and projects identified. After about two years, the company has identified more than fifty projects in all areas of business. It takes a while to grasp the concept of bottom-line results, given past experience with quality initiatives, which teach that results come over time. Unlike at large companies, where savings from each project are expected to be about $250,000, at smaller companies, savings could be targeted at about $100,000. Now, for the company management, ISO certification or number of employees with training certificates are less important than the actual bottom-line results. Becoming a better and more responsive supplier to customers is more important for growing the business. Results matter to smaller companies much more than at larger companies. Effort and investment, as well as results, are more visible within a short time.

To ensure progress and results, the management establishes a management review consisting of operational as well as financial results. The savings that are accounted due to Six Sigma projects are those that have already been realized with the process improvement effort. Companies can save a lot of money by implementing aggressive process improvement activities that use the Six Sigma methodology. Just as cost of poor quality could be between 10 and 25 percent, similarly, savings due to the Six Sigma initiative could be as high as 10 to 15 percent of sales dollars. Owners would love to have that much money in their pockets, or use it to grow their business.

Another scenario occurs when a smaller company gets a new CEO or president who has experienced Six Sigma success in his or her previous job. After assessing the state of the new organization, the CEO decides to build a Six Sigma organization to turn the business around or increase the bottom-line performance. Once on the Six Sigma journey, the commitment must be a long-term and sustainable one. To ensure that, cost and benefits analysis must be performed and opportunities identified at the company, under current and future circumstances. A commitment to extensive training must be made to change the mindset and learn the new definition of quality, a change from building to specifications to building to targets.

Initially, the focus at smaller companies can be to reduce cost or waste in the system. Considering competitive pressure on prices and rising costs of material and labor, it is an imperative that cost of operations be reduced. One way to reap quick benefits is to reduce scrap. Such a project would be applicable at many smaller companies, and would make potential benefits quickly visible. Savings could start from $500,000 and move upwards. For smaller companies, Six Sigma had better be about making more money—otherwise, it stinks.

EXAMPLES OF SUCCESSFUL APPLICATION OF DMAIC METHODOLOGY

Example 1: Chemical Plant Bottleneck

Define. Distillation tower has internal damage, limiting production rates. Next outage is scheduled in one year. If outage is taken now to repair damage, we will still have to take outage in one year because of parts delivery for other essential projects.

Measure. At anything over 85 percent of capacity, the distillation tower will not perform. With six months of effort, operations engineers and process engineering could find no solution other than to take an early outage. Anything less than 100% capacity is considered a defect.

Analyze. Identified key operating variables, established allowable ranges for each, and conducted a designed experiment.

Improve. A single set of conditions allowed operations at 102 percent of capacity without problems. At that level, another part of the plant became the bottleneck. The increased capacity until the scheduled outage was worth $6 million.

Control. All shift operators were trained for new conditions, and the operation procedures were modified.

Example 2: Water Treatment

Define. Water-treating unit in fifteen years had never been able to handle the nameplate capacity. Treatment chemical costs were higher than other types of treatment units.

Measure. Confirmed actual flow rate through the system, versus nameplate flow rate.

Analyze. Measured system evaluation and found many measurements that were off by over 100 percent. Hourly operations identified key variables in the operation of the unit and the acceptable range of each. Conducted three different designed experiments.

Improve. Corrected the measurement problems. Found set of operating variables that produced 107 percent of nameplate capacity at higher quality with lower chemical use. Chemical use reduced by $180 K per year.

Control. Hourly operations trained, procedures modified, process to check measurement instituted. Introduced model for changes in inlet water conditions.

Example 3: Power Distribution Reliability

Define. Large chemical site had significant losses due to power outages.

Measure. Dollar value determined for each failure and the total. Each failure was assigned to a major component.

Analyze. Mapped the entire system by major component, and identified failure rates for each major component. Found areas with projects scheduled that were very unlikely to fail, and thus the projects would add nothing to overall reliability. Other components that had a high likelihood of causing an outage were being ignored.

Improve. Developed a plan for each component, depending upon failure mode and frequency for that component. Made a tenfold reduction in the dollar losses due to power failures on site.

Control. Track each major component and modify action plan based on failure mode, if needed. System shared with other locations.

13

ENHANCING SIX SIGMA TO DELIVER HIGHEST PERFORMANCE AT LOWEST COST

Even though Motorola saved over $2 billion in operational costs through Six Sigma, it is struggling today. Polaroid, another major corporation that derived significant benefits from Six Sigma, has filed for bankruptcy. So, why is the promise of significant benefits not delivering long-term results in some organizations? The authors' review of this issue has led to an enhanced approach that provides sustained performance improvement by combining a regular strategic planning cycle with the use of Six Sigma as a methodology to achieve the strategic objectives of the organization.

THE TRADITIONAL APPROACH TO SIX SIGMA IS A BOTTOM-UP APPROACH

Since their innovation by the late Bill Smith of Motorola (Schaumburg, IL), Six Sigma concepts have been widely utilized in electronics and other industries. Having worked with Bill Smith in 1987–88, we can safely say that Bill probably never dreamed that one day his idea of measuring products and services would become an industry in itself. He would be proud of the publicity his innovative thinking is receiving. But, at the same time, he would be amazed to know how the scope of Six Sigma has changed over time. Corporations such as GE, Motorola, Allied Signal, ABB, DuPont, Dow Chemical, and others have used Six Sigma as the primary tool to achieve their profitability improvement objectives.

Six Sigma is a measure of the effectiveness with which products and services are delivered. It encompasses the entire supply chain, ensuring quality at every step of the process. Higher sigma means better quality, and lower sigma means poorer quality of a product or service. Average American companies operate at a quality rating between three sigma and four sigma. Six Sigma is a much higher quality level, representing 3.4 errors per million opportunities. It is an approach that measures current quality level, and if it is at an unacceptable mark, a team is tasked to address it and raise the quality level to Six Sigma by reducing the errors and defects.

The key elements in traditional Six Sigma initiatives have included:

- Aggressive goal setting
- Highly-trained operational and executive management
- Effective quality management review
- A standardized measurement system
- Use of a "small wins" concept to validate the Six Sigma measurement model
- Graphical reporting of performance against goals
- Inspiring leadership

Six Sigma literature published to date has been largely a collection of quality-related statistical tools that have been in existence for a long time. However, as they have been compiled in the age when Six Sigma is gaining greater acceptance, these quality tools are now considered Six Sigma tools. Conventional practices and improvement techniques are being linked to Six Sigma, as with Six Sigma Lean Management, Six Sigma Designs, and Six Sigma Kaizen. Six Sigma Black Belt certification has become even more expensive than earning a graduate degree at most state colleges. Corporations must commit to a significant investment to realize the benefits.

Because the investment is high, realization of benefits is essential, or else the initiative will not create value for the corporation. Too rarely is the question asked, "What will be the rate of return on this investment?" As with any corporate initiative that competes for corporate resources, there must be a focus on initiatives that achieve highest performance at lowest cost. If not, we are not optimizing the value of the corporation.

IMPRESSIVE RESULTS HAVE BEEN ACHIEVED

There are some great success stories. Motorola grew dramatically between 1987 and 1992, the first five years of Six Sigma. During this time, the company's sales doubled, profit margins improved, and its reputation soared. Those who worked in process improvement at Motorola did not just throw dollars at Six Sigma to achieve this dramatic improvement. Instead, simple practices, such as strong corporate leadership, good project management, teamwork, an excellent reporting system, communication, and equitable rewards, helped achieve dramatic quality improvement at Motorola.

Motorola reported about $2 billion savings in manufacturing operations during the first five years it implemented Six Sigma. Similar gains were also realized in the nonmanufacturing areas of the company. In addition, other companies have reported dramatic savings from utilizing the Six Sigma methodology. Below are some very impressive quotes from the leaders of some early adopters about their experience. These and others can be found in *Six Sigma: The Breakthrough Management Strategy*, by Mikel Harry.

- General Electric (Jack Welch): "The initiative started in 1996 delivered $300 million to the bottom line in 1997 and $600 million in 1998."
- Allied Signal (Larry Bossidy): "The initiative introduced in 1994 has generated savings of over 2 billion dollars and saved the company from bankruptcy."
- Raytheon (Daniel Burnham): "The initiative introduced in 1998 is expected to generate savings of over 1 billion dollars annually starting 2001."
- Polaroid (Joseph Casabula): "The initiative adds six percent to the bottom line each year."

As the adoption of the Six Sigma methodology gained broader acceptance within U.S. corporations, norms for expected levels of success have developed. Typically, a Black Belt (a professional with leadership training in Six Sigma) can return about $250,000 per project to the bottom line, and complete about four projects per year. So, if a company has trained about a thousand Black Belts, they can expect additional PBT (profit before taxes) of about $1 billion during the first year. Of course this assumes that as part of the project identification process enough projects of this size are identified. It does no good to have a thousand Black Belts if there are not good projects for them to address. Project identification and selection should align with the organization's strategic plan, and is a management responsibility.

But There Are Shortcomings in the Current Approach . . .

Sustaining this remarkable performance over long periods has been a problem. First, such programs have not been linked to the corporate strategy, and hence lose sight of their objective. Motorola, the pioneer of this methodology, itself ran into performance problems in later years. A similar fate has been met by a number of other companies, such as Polaroid. As more companies commit to deploying Six Sigma, they must ensure that there is a link to their strategic objectives.

Some overzealous communication of success stories and the strong and marketable leadership of a few organizations have led Six Sigma to a path that includes a cowboy mentality to quality, increasing the cost of quality improvement and causing long-term damage to the quality profession. The process-oriented performance improvement has become a program-driven strategy. Experience tells us that programs are not sustainable. When you have a process that helps to deliver the organization's strategic objectives in a more efficient and predictable fashion, thus adding value, that process becomes an essential part of how an organization does its work.

Key lessons we have learned from failed experiences include:

- Improvements in product/service quality need to be linked to corporate strategy.
- Improvements in quality must be achieved in a cost-effective manner. Only if benefits outweigh the cost should any initiative or project be undertaken.
- There must be a process for sustaining improvements.

The Improved Methodology

Given these limitations, we have to address them through our methodology. This approach integrates some key elements from the Balanced Scorecard (developed by Robert Kaplan, a professor at Harvard Business School) that address these issues. The top-down approach offered by the Balanced Scorecard is driven by the corporate strategy. When that is combined with the bottom-up tactical approach of Six Sigma, it delivers powerful results.

Strategic Alignment

Strategic alignment is the most critical factor for sustaining the Six Sigma initiative. Table 13.1 identifies potential problem areas resulting from misalignment of business objectives.

The implementation team can never lose sight of the strategic objective they are trying to achieve.

Cost Benefit Analysis

The next most critical factor for sustaining the Six Sigma initiative is a convincing value proposition. To evaluate the Six Sigma methodology for their company, management must consider both the benefits and costs of implementing the methodology. To determine the potential benefits to be realized from the Six Sigma methodology, the management must first understand the cost of their operations. If a detailed breakdown is not available, they can

Table 13.1 Misalignment of Business Objectives—Potential Problems

Strategy	Customer	Quality	Processes	Financial
Strategic objectives not clearly specified	Customers dissatisfied	High error rates or exception processing	Processes that are broken and inefficient	Declining or stagnating revenue or profits
Operations not clearly linked to strategic objectives	Market share declining	Issues with supplier/ partner performance	Departmental focus overriding the enterprise's effectiveness	Need to increase shareholder value
Unsure if current initiatives will deliver expected results	Customer responsiveness low	Low yield rates	Effectiveness and efficiency not being measured	Declining EPS
	High level of returns or disputes		Multiple reentry of data	

use industry benchmarks. The cost of poor quality, in terms of internal and external failures and cost of inspection, test, or verification, must be computed. Unless accurate data about the company's current performance are known, all initiatives will appear questionable. In our experience, the cost of poor quality is usually grossly understated. This is especially true when the cost of external failures (failures that reach the customer) is considered. As a recent example, consider the cost of tire failure for Ford and Firestone.

Additional uncertainty about committing to a Six Sigma initiative comes from several unknowns, such as the level of effort involved, amount of financial commitment, probability of success, and fear of failure. One way to overcome these uncertainties is to establish specific, measurable, attainable, realistic, and tangible (SMART) goals. The additional intangible benefits (in addition to hard savings) that will be achieved—such as awareness of strategic objectives, cultural change, and focus on quality—will help guide the decision making.

For management to commit to a Six Sigma initiative, certain questions must be answered:

- How can the Six Sigma initiative benefit the company, in the short term and in the long term?
- How much will it cost?
- Do the benefits justify the cost?

Sources of Benefit

The hard benefits realized by the company include tangible items such as reduced waste, lower reprocessing and warranty costs, improved profitability due to increased margins, lower investment in working capital, and so forth. In addition, there are softer benefits to be gained through an engrained culture focused on Customer's Critical Criteria, leadership development, employee alignment, and quality.

Cost Elements

Recently, Polaroid filed for bankruptcy, Motorola struggled for survival, and some quality gurus questioned the Six Sigma methodology. It is only natural to look into the details of implementing Six Sigma. The cost elements include planning, organization, training, project implementation, communication, and recognition.

Table 13.2 Six Sigma Estimated Costs

Activity	% of Total Cost
Planning	5%
Organization	10%
Training	50%
Implementation	30%
Communication	1%
Reward and Recognition	4%

Table 13.2 provides an estimated cost breakdown for implementing Six Sigma in an organization. These values are only approximate and can vary.

Currently, the breakthrough strategy requires a significant expense initially for corporate-wide executive, Champion, Green Belt and Black Belt training. Projects are selected and interwoven between training modules. Companies that have committed to significant training corporate-wide are struggling to identify projects to achieve any return on investment. The goal of Six Sigma is to realize highest quality at lowest cost. The high cost of implementing Six Sigma will eventually add to the cost of operations if the right projects are not identified. As we have shown, training of Black Belts and Green Belts should not start until there are sufficient high-value projects to keep all of the people productively engaged. Training can be very expensive, not only for the actual delivery of the material but also in the dollar value of time of the people receiving the training. With the approach we have developed, this trap is avoided.

Economic Viability

The following are some key factors that must be considered in order for a company to estimate the benefits and costs of Six Sigma methodology:

- Number of employees
- Annual revenue growth ($)
- Cost of poor quality ($)
- Cost of implementing Six Sigma (internal and external)
- Minimum number of projects to be identified to break even
- Minimum number of projects to be identified for 100% return on investment

Use of activity-based costing (utilizing cost drivers) is the preferred way to measure actual costs, compared to traditional costing systems in which significant portions of the overhead costs are allocated on the basis of revenue or head-count. Further, when costs and benefits are derived over a period of time, use of the discounted cash flow method to determine the return on investment is preferred. When doing project evaluations, the hard dollar savings are separated from the soft dollar and avoided cost savings. Unless projects demonstrate high probability of actual contribution to PBT, they should be culled from the list. These projects may have other redeeming qualities, but do not dilute the contribution of Six Sigma by including them in the Six Sigma project list.

Once the economic viability of a Six Sigma initiative has been established, it must be given the highest priority, and management must invest in the required resources.

Our Approach

To address the limitations discussed earlier, the new approach must incorporate a review of strategic alignment and an analysis of benefits and costs involved (Figure 13.1).

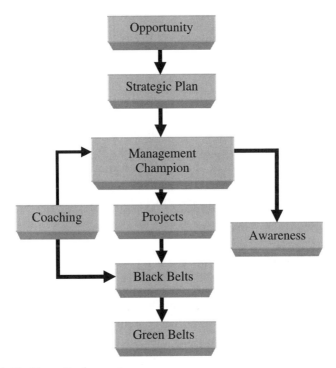

Figure 13.1 Six Sigma Deployment.

This process is not a single, one-time effort. On the regular strategic planning cycle, usually at least once per year, the opportunity and strategic plan are reevaluated. Potential projects are being added, assigned, and removed as an ongoing process. Black Belts are not limited to that role for a career; typically they move to more responsible positions, frequently in management. This requires a review of the plan to train Black Belts' replacements. Green Belts are typically project-specific, and people assigned to projects are trained as needed. New employees are always being added to grow a viable organization, and ensuring that they have an awareness-level understanding of Six Sigma is part of the new employee development plan.

Review the Business Case

As discussed earlier, this is a critical step that is frequently overlooked. There is usually a cost involved in improving quality—test equipment, inspection, improved process, higher quality materials, and so on. In order to ensure that the quality improvement initiative will add value to the organization, it is important to compare the cost with the benefits to be derived. Only if benefits outweigh the cost, should the initiative be pursued. The benefits typically include revenue growth through increased sales (market share), capability to charge a higher price due to superior product or service, and lower cost due to less scrap and fewer errors, defects, or returns.

Define Business Goals

The key to executing any strategy is the ability to communicate it clearly. Articulate the strategic goals that need to be achieved. The Balanced Scorecard suggests that these goals cover four key areas: finance, customer, process, and learning. Using this framework, management must identify strategic objectives in each of these areas. Although the majority of companies continue to give financial objectives much greater weight than any of the other areas, the others are extremely important for sustained financial success.

Verify Strategic Alignment

Align the operational initiatives to the strategic objectives; they are "what" we must do operationally to accomplish the strategic objectives. The role of Six Sigma is defined here—whether it is implemented with the intent to reduce cost (increase profits), enhance customer satisfaction, or optimize process efficiency, Six Sigma methodology can play a role and help achieve desired results. It is important to clarify the objectives and the initiative that will help achieve them. Six Sigma is a very desirable tool to achieve many quality-related objectives.

Measure and Analyze

Having screened the initiatives to include only the ones that show the promise of increasing shareholder value, the next step is to perform detailed analysis of the problem area. This leverages the robust analytical framework offered by Six Sigma tools to achieve breakthrough results. Additional activities include establishing short- and long-term performance measures and improvement targets. The analysis focuses on (1) determining what is the current effective yield of the product or service, and (2) if it is less than Six Sigma standard, determining what it would take to achieve that standard.

A cornerstone of the Six Sigma methodology has been to aim for a dramatic improvement, an improvement that is visible, meaningful, and profitable. Merely setting an incremental improvement goal is not what Six Sigma is about; it will drive for a 70 percent improvement rather than a 10 percent improvement. Setting a goal of 70 percent for improvement requires a totally different approach. The 10 percent improvement goal is not perceived as a challenging goal—it can be realized with minor adjustments in the process. Even though a 70 percent improvement goal first appears impossible, once thought through with a commitment to achieve the target, it starts becoming a reality. The real challenge for management is to expect and sustain goals for dramatic improvement.

Plan and Manage Transition

The detailed analysis not only provides a definition of the current state, but also defines the desired future state. Comprehensive planning is required to transition the organization from the present situation to the future state. It also involves identification of resources required, and consideration of how change will be introduced with the least disruption to current operations. It involves detailed evaluation of changes required across three key dimensions: people, process, and technology.

Enhance and Sustain Performance

The methodology focuses on organizational elements that will help sustain the improved performance over a period of time. Many initiatives suffer as they become one-time improvement efforts. The performance tends to return to the previous levels once the focus is removed. Existence of processes to accurately measure and report progress in a timely manner is critical to the initiative's success. In addition, it is also important to provide incentives to the implementation team to achieve and sustain the improved results. Finally, the process needs to be periodically reviewed to address any inefficiency and leverage any learning from past experiences.

Organizations following this detailed process are certain to achieve breakthrough performance improvement and a high return on their investment.

CONCLUSIONS

The current trend in utilizing Six Sigma as a corporate methodology has corporate executives committing to a significant investment in resources without ensuring strategic alignment and determining appropriate targets for return on investment. As a result, the investment in the Six Sigma initiative is exposed to a high risk and probability of achieving low return on the investment. The authors believe that the executive management must plan such an initiative well before committing significant resources to it. More importantly, such a road map will set the right direction for the Six Sigma journey and ensure that the corporation receives sustained benefits over a long period.

14

SIX SIGMA METHODOLOGY— THE BENEFITS OF A STRATEGIC APPROACH

OVERVIEW

Even with all of its success, some view Six Sigma as a fad and an expensive program that will fade as newer, more trendy programs come forward. These naysayers usually cite examples in which Six Sigma has not resulted in the desired performance improvement, or the implementing company has lost market share or even declared bankruptcy. Motorola's loss of market share in the cellular phone market and its Iridium satellite phone business, as well as Polaroid's bankruptcy and Whirlpool's lack of financial results, is widely documented. These cases are troubling, since Motorola was the inventor of Six Sigma, and only a few years ago Polaroid was touting its Six Sigma program as its savior. However, as with most things, this is too simplistic an analysis of what Six Sigma is and where it fits into an overall strategic management system. Properly deployed in a total systems approach, Six Sigma becomes one of the most powerful tools in the strategy of a corporation.

In many cases, Six Sigma is implemented as a stand-alone program that will be the savior of the corporation. This approach is usually taken when there is a direct threat to the organization's existence, and it focuses on manufacturing. The rest of the company often views manufacturing as the cause of their problems. At this point, the pressure to reduce cost is overwhelming, since everyone, especially in sales and engineering, is convinced that lowering price is the only way to stay competitive. Consultants are hired, slogans are developed and disseminated, training is completed, and programs are launched. However, most of these programs are just methods of diverting attention from key problems, and are only buying time. Since defects are reduced and yields improve for a short while, victory is declared, and the company continues on its path of decline. Six Sigma is then blamed for not meeting the expectations of the corporation and its stakeholders.

This scenario is understandable since most executives are taught to break a problem down into component parts and develop solutions that focus on a single issue. This analytical approach allows for solutions to be developed that are optimized at a component level in the organization but don't fit well with the overall strategy in the organization.

Under this type of thinking, if you were to set out to build the world's finest car, you would go to the world's leading designers and manufacturers of components that make up a car and tell them to simply deliver the best part they can make. Of course you wouldn't do this, since there is absolutely no chance that the components would fit together to make a functioning automobile. You would define specifications and interfaces that would define the car as an integrated system.

This is the fundamental reason the dissatisfaction of organizations that claim to have deployed Six Sigma yet have not realized the desired improvement in performance. Six Sigma cannot save an organization that has a flawed strategy or does not understand Customer's Critical Criteria. Building the best analog cell phone systems when customers are demanding digital systems, or ignoring a major shift in consumer demand to digital cameras, will not be offset by any Six Sigma program, no matter how well it's implemented.

SYSTEMS THINKING IS REQUIRED

To make Six Sigma successful requires a new way of thinking. As Albert Einstein said, "Problems that are created by our current level of thinking can't be solved by that same level of thinking." To quote *Inc.* magazine, from an article on W. Edwards Deming: "Deming's perspective reflects a more Eastern habit of mind. In the Western world, we seek to affix blame and reward to individuals. Deming teaches that we must reject that lens and look instead at the system in which the individuals operate."

What both of these great thinkers are telling us is that we need to take a higher-level view of Six Sigma and understand that it is an integral part of an overall organizational system. In addition, we need to recognize that organizations are made up of human beings whose lives and work follow systems. Once we understand these concepts, Six Sigma can deliver outstanding performance over a long period of time.

Systems Within Systems

An individual, a team, a division, and even the outside world are integrated and interrelated parts of an organizational system. This means that every one of its parts affects the performance of the organization as a whole.

However, the issue is more complex than that. Each level of a typical system interfaces or collides with other systems. Individuals must work with other individuals, groups, or departments. Departments must interact with the whole organization, and the organization must interface with the outside world. Managing these interactions is critical in an organizationally encompassing program such as Six Sigma. Specific leadership skills are needed at each level and interaction, to ensure readiness and proper implementation of any Six Sigma program.

Begin with the End in Mind

Generally, we analyze processes linearly and define them as an input-transformation-output model. We usually define the input or current state and develop a SWOT (strengths, weakness, opportunity, and threats) analysis. This is where most Six Sigma programs start. Management knows that something is wrong with the current performance of the company and looks to improve it. A program is initiated and implemented, and the outputs (hopefully improvements) are measured. This model works fine when we are doing specific problem solving and are reacting to situations that need to be addressed immediately. This model, however, is lacking when using Six Sigma to attack long-term problems that encompass all levels of the organization. This is where the concept of systems thinking comes into play.

As Stephen Covey said in his book *The 7 Habits of Highly Effective People*, "Begin with the end in mind." We need to start thinking backwards. First define what the goals are for the organization (especially as they relate to the customer in our future ideal world), while keeping in mind the environment, *as it will exist in the future*. We can then create the measurement criteria that describe the specific Key Success Measures that define success. These clear, concise, and bounded measurement criteria will be used to evaluate all the strategies that will be implemented in the company.

Once this is complete, we can look at Six Sigma in the light of these criteria and decide if it is truly a strategic initiative that needs to be implemented. Doing this and communicating this analysis to all the stakeholders in the organization, while getting their input, is a powerful tool in obtaining general acceptance of the need for Six Sigma. Six Sigma can then be used as a means to align all the processes in the organization to meet the overall strategy.

At each stage in the systems approach to Six Sigma, continually ask some very basic questions that will help to make sure things are on track:

A. Where do we want to be? (What are our ends, outcomes, purposes, goals, destination, vision?) This is the basic question of systems thinking and needs to be future-oriented and focused on the Customer's Critical Criteria.

B. How will we know when we get there? (What quantifiable feedback system will link to the Customer's Critical Criteria?) This will define the criteria to measure the success of *every* Six Sigma initiative throughout the organization.

C. Where are we now? (What are today's issues and problems?) This is the place to do a SWOT analysis and evaluate the current environment.

D. How do we get there? (How do we close the gap from C to A?) This is where Six Sigma comes in and aligns all the processes involved within the organization. Other human attunement processes, such as reward and recognition programs, leadership development, and employee evaluation systems are involved.

E. What will/may change in the future environment? (What's happening that could change our plans?) This needs to be accomplished on a consistent and definable basis. It is also the beginning of the program.

Change Cycles

Strategies that help people "buy into" and "stay into" a Six Sigma implementation must also be developed. As Six Sigma begins to be introduced to the organization, each person will go through a change cycle. At first, people will deny that any change is necessary and will be shocked that management believes that they or their operations need to change. After a while they will become angry or depressed when they begin to see how the changes affect them personally. Both of these feelings will happen to everyone in the organization at some time, although seldom all at the same time. There is no guarantee that these feelings will mitigate, and it's management's responsibility to help employees get through this period. If management fails, it's entirely possible that the workforce will again slide into shock and denial.

If the leadership is successful in listening to the employees' concerns, empathizing with their feelings, and communicating why Six Sigma is necessary, each individual in the organization has the opportunity to accept the changes. At this point management should show why Six Sigma would be beneficial to them and their groups, and get individuals to participate in the change process directly. Some employees are now ready to embrace Six Sigma as an integral part of their jobs and careers, and to build a more robust organization that performs at a higher level. Remember, it is very unlikely that all employees will progress through these stages at the same pace. Some will be ahead of the norm, and others lagging.

SUCCESSFUL IMPLEMENTATION

Looking at Six Sigma through the lens of a systems approach, we can begin to understand why GE has been so successful in its implementation. In their 2000 annual report, we see that Six Sigma is one of four strategies for success (included with Globalization, Services, and Digitization) and is used as the fundamental way in which all GE employees run their processes. These strategies are based on the values: A Learning Company, Integrity, Relishing Change, Focus on the Customer, Leadership, Self Confidence, Simplicity and Speed, People, and Informality. There is a fundamental emphasis on customers, employees, and the practices that allow them to succeed. GE's practice of ranking and removing people/managers who are not performing is legendary. Each year the environment and personnel are reviewed, and new programs and strategies are brought to the forefront and implemented. This is the basis of GE's success. Six Sigma remains a core strategy throughout.

Making Six Sigma Successful in Your Organization

Six Sigma initiatives focus on improving profitability and customer satisfaction, rather than simply improving quality. Six Sigma initiatives are for companies that want to be profitable and achieve world-class performance.

A typical company, managed by executives looking at averages and satisfactory performance, will achieve average results.

Where do we want to be? This is the basic question of systems thinking, and it needs to be future-oriented and focused on the customer. Before starting a project or implementing Six Sigma company-wide, the results must be clearly visualized. The intent of the initiative must be established and communicated. Goal setting that requires a significant rate of improvement in business performance leads to more systemic improvement, rather than an incremental rate of improvement that only requires a little manipulation of the business processes.

How will we know when we get there? This requires the criteria to measure the success of *every* Six Sigma initiative throughout the organization. To overcome bottlenecks to higher profitability, a company must establish a strategy including a vision, mission, vision elements, key performance indicators, objectives, goals, and plans. To benefit from Six Sigma implementation, all must recognize the value of improving profitability, have passion to achieve improved results, and become committed to improving customer satisfaction.

Identifying an area or division for piloting the Six Sigma initiative is a good way to develop a successful program. Having in-house success stories can be a great way to gain interest in other departments, divisions, and management sectors. Ultimately, employees must be able to see the improvement, share the benefits, and actively participate in accelerating the improvement.

Where are we now? A Six Sigma project, or the corporate-wide Six Sigma initiative, begins with establishing a baseline. The base can be established by collecting data under normal operating conditions. Once the current level of performance is determined, and the expected performance levels can be established, an organizational baseline must also be established. This is the time to do strength, weakness, opportunity, and threats (SWOT) analysis. The organizational baseline includes understanding organizational structure, policy, and procedures, informal structure (politics), human capital and liability, and financial impact and investment resources.

To implement the Six Sigma methodology, the plan must also include the education of executives in systems thinking, Six Sigma methodology, expected challenges, and how to overcome these challenges. Organizational leaders must understand the concepts, steps, requirements, expectations, and leadership skills necessary to actively participate and contribute to success. Employees should never have reason to doubt a leader's priority to improve profitability and meet or exceed Customer's Critical Criteria. All employees, including the senior managers, must understand the consequences associated with applying or ignoring Six Sigma methodologies. The entire company must have a common goal, a plan, and common priorities to make the Six Sigma initiative successful. The team must be like an orchestra playing and enjoying its music together.

How do we get there? Six Sigma aligns all the processes involved in the organization. Other human attunement processes, such as reward and recognition programs, leadership development, and employee evaluation systems, are included. Employees receive the awareness training to ensure successful implementation. Those who can, apply statistics and new tools. Those who are keen observers and experimenters, have a curiosity to investigate/solve problems, and like to be trainers/facilitators should be selected for additional training to achieve Black Belt or Green Belt levels of competency. Black Belts will eventually become project team leaders. The successful Black Belt practitioners must be preferred for further development into leadership positions. Teams that have produced exceptional results must be recognized at the highest level, not only to reward them, but also to encourage more contributions from future Black Belts and their teams.

The following action items are needed to implement the Six Sigma initiative:

- Develop or review the organizational strategic plan:
  - Conduct a "Plan to Plan" session with top management.
  - Review the environment outside the corporate level.
  - Establish or reaffirm values and future business objectives.
  - Establish profitability objectives.
  - Perform financial analysis to understand profitability, COPQ, and key contributors.
  - Measure contributors to profitability.
  - Establish performance and organizational baseline.
- Conduct a "Plan to Implement" session with key personnel.
  - Recruit a firm for Six Sigma training and implementation guidance.
  - Select a pilot project or area for "small wins."
  - Define projects and develop plans to realize improvement.
  - Conduct executive, Champion, Black Belt, and Green Belt training.
- Solve problems and develop solutions to reduce waste.
  - Monitor progress of projects and provide support as needed.
  - Celebrate and publicize successes.

    ○ Learn lessons from small wins and optimize approach.
    ○ Institutionalize the Six Sigma initiative company-wide.

What will/may change in the future environment? To prevent wasteful planning, one must consider future as well as current events that could change plans. This needs to be done on a consistent and definable basis. Anticipating changes in the external environment, the organization, and internal capabilities is fundamental to any Six Sigma initiative. The best way to anticipate changes is to understand all of the surroundings and take a proactive, holistic approach. Various combinations of difference variables or conditions could lead to an infinite number of problematic conditions.

The pioneers of Six Sigma and systems thinking have understood that we do not control many conditions. Understanding where Six Sigma fits within the organization's environment gives a clue as to where changes may be necessary. This allows "outside the box" thinking that expands the thought process to arrive at creative and innovative solutions to problems. Just remember the maxim "It is a bad plan that admits of no modification."

In the absence of systems thinking, or when Six Sigma concepts are applied in a limited scope, the following symptoms are quite commonly observed:

- Six Sigma is viewed as a quality improvement tool like others that may not have worked in the past.
- Customer's Critical Criteria are not understood and are not a focus of attention throughout the organization.
- Quality focus and objectives are not clearly defined or are poorly communicated.
- Executives continually think that quality has nothing to do with business and profitability.
- Measurements (levels and trends) to track operations performance, including customer satisfaction, reject rates, rolled yield, COPQ, design effectiveness, cycle time, inventory levels, employee skills development, and financial performance, are not in place.
- Centralized decision making (executives making the decisions) is the norm.
- Executives are busy fighting fires, making an effort to "look busy," and hassling employees.
- Employees are afraid of management—reluctant to take the initiative to improve performance—and feel that no one is listening to concerns.

The benefits of implementing Six Sigma are too attractive to overlook. Increased customer loyalty, more revenues, improved operating margins, higher returns, and increased earnings are all demonstrated outcomes. The question that must be answered is not whether to implement Six Sigma; instead, the question is how to implement Six Sigma to sustain better business performance as part of a whole organizational system.

15

SIX SIGMA AND QUALITY SYSTEMS

Six Sigma, a customer-oriented methodology to maximize an organization's profitability while producing the highest quality product, consists of two key components: methodology and measurements. Methodology is customer-centric and focuses on improving processes to achieve desired results. Measurements are in place to ensure that the methodology works—easier said than done! In the current Six Sigma environment, methodology has become a project-based institutionalization of process improvement. Considering the successes and failures of well-known corporations, it is clear that leadership, goal setting, and measurement are the key factors that make Six Sigma work. Leadership emphasizes the cultural aspects of the methodology, goal setting lays the groundwork for aggressive improvement, and measurement is a verification of the improvement.

To ensure that desired improvements are achieved, measurements for profitability, quality, timeliness, cost, and customer satisfaction must be in place. The first step in implementing Six Sigma is to assess the state of the business (methods, material, machines, and employees), to identify strengths and opportunities, and to establish a baseline for key measurements. For areas with opportunity for improvement, establish improvement objectives prior to implementation of Six Sigma methodology.

The first challenge of the process is to identify the aspects of the business that should be measured. What is the simplest way to determine important aspects? Ask why. If the quality of a product or service is important, ask *why* quality is important to the company. The following questions may help your company to explore and identify desired measurements:

- Business objectives. Why is a product, process, or business characteristic important to the business?
- Success factors. What goals should be achieved?
- Input measurements. What is needed to achieve these goals?
- Process measurements. How are these goals achieved?
- Output measurements. How is achievement of those goals determined?

By asking such questions, a company can identify process measurements to ensure excellence. For service operations, answering these questions would identify process measurements—helping a company understand units, improvement opportunities, and the concept of defects per million opportunities (DPMO) to achieve Six-Sigma-level performance.

Six Sigma measurements identify customer-critical characteristics, evaluating performance at any given process step, or, for a product, it calculates process capability to determine the probability of success and improvement in terms of sigma level. For any given sigma level, there is an unacceptable level of performance that can adversely affect profitability. The main purpose of the measurement system is to quantify performance and its impact on profitability. If performance significantly affects profitability adversely, a project can be implemented to solve the problem.

To determine performance, the process defect rate is normalized per the process output, called a unit. A unit is a discrete quantity of output that can be counted, verified, and measured. In an assembly process, each circuit board assembled could be a unit. Similarly, each board coming from the wave-soldering or reflow process is a unit. A unit is used to estimate the quality of the process output in terms of defects per unit, percent yield, or the first pass yield.

Defects per unit (DPU) is defined as a ratio of the total number of defects observed in the inspected or verified units over the total number of units processed or built.

$$DPU = \text{Total number of defects/Total number of units verified}$$

Know the difference between a defect and a defective unit. One defective unit may contain many defects—the goal is to count all defects.

When unacceptable DPU exists, the problem may be addressed by understanding which process step is involved, where the problem occurs in the process, and the root cause, which may be the result of any part or process. Plenty of opportunities for things to go wrong exist in the manufacturing process. Such opportunities become critical when comparing products of various complexities and solving process problems.

DPMO is a measurement that normalizes the reject rate based on opportunities instead of units:

$$DPMO = \frac{\text{Total number of defects} \times 1{,}000{,}000}{\text{Total number of units verified} \times \text{Average number of opportunities in a unit}}$$

To understand the relationship between DPU and DPMO, the formula above can be restated as:

$$DPMO = \frac{DPU \times 1{,}000{,}000}{\text{Average number of opportunities in a unit}}$$

Avoid misuse of the formula. Improving DPMO by increasing the number of opportunities is unacceptable. Instead, the objective must be to reduce the number of opportunities by reducing part counts and process steps. Quality is improved by reducing defects—resulting in an actual improvement, not just an arithmetic improvement.

Units, or opportunities, and defects, or errors, are countable items—the formulae above work in cases of assembly or service processes. In cases of continuous processes, where variation is measured, the probabilities of producing defects should be applied to the formula. Probabilities consider expected shifts in the process—established at 1.5 sigma. Probabilities are calculated and transformed in DPMO measurement.

The main element of the Six Sigma methodology is to identify strategic opportunities for improvement and to improve processes dramatically. Imagine if one were asked to improve process performance by 10 percent per year. One could easily think of identifying a few things in the process to change, and plan to achieve the 10 percent improvement. However, if one is asked to improve the process performance by 70 percent per year, one has to think hard. Such a high rate of improvement requires innovative thinking and practically reengineering every year. To achieve such a dramatic rate of improvement, strategic planning, extensive training in statistical project management, and strategic tools are needed. Then, to plan to improve an organization's performance continually, management must establish a direction for continual improvement. The new ISO 9000 and similar standards have incorporated requirements for continual improvement.

Six Sigma methodology is a great way to drive continual improvement in an organization, and quality system standards are a great way to sustain such an improvement. The quality system is audited by a third party such as Underwriters Laboratories, Lloyds, NSF, KPMG, TUV, and dozens more. Once the Six Sigma methodology becomes part of the quality system, it is audited at the grass-root level to ensure effective implementation. The quality system requirements for training will help track training objectives and effectiveness; provision of products and service requirements will ensure control of process conditions; continual improvement requirements can accelerate the results; and management review will maintain the drive to achieve success of Six Sigma projects.

The following sections describe the ISO 9001 quality system and its elements, which can facilitate the Six Sigma methodology, and other standards such as AS-9100, TL-9000, QS-9000, ISO 16949, and ISO 14001.

ISO 9001:2000

ISO 9000 standards have been in existence for about thirteen years. The number of companies implementing the quality system and achieving registration has been growing since its release. Several hundreds of thousands of organizations have achieved ISO 9000 registration and are benefiting from it. Now the standards have been changed to achieve even better quality.

Surveys have found that the most commonly stated benefits of implementing ISO 9000 quality systems include the following:

- Improved product quality
- Consistency, standardization, and repeatability
- Increased business
- Increased customer confidence
- Better management and less confusion in the plant

Besides all the benefits of implementing an effective quality system, there are opportunities for improvement in the current requirements of the standards.

Some of the industry-wide concerns include the following:

- There is a lot of documentation or paperwork.
- There is no change in doing business.
- Management does not care.
- Quality has not improved.

The standards committee heard all the issues and revised the standard. They devised a new standard that has been reviewed by many companies, consultants, and registrars. The result of the committee's efforts: the ISO 9001:2000 version was released in December 2000.

The new standards reemphasize the process approach and facts-based decision making. The process approach consists of four major elements: Management Responsibility, Resource Management, Product Realization, and Measurement, Analysis, and Improvement. The model recognizes the significant role played by customers in defining requirements as inputs and providing feedback to validate that the requirements have been met.

The previously published three models of the ISO 9000 standards have been merged into the new ISO 9001:2000 standard. Also, the terminology of "subcontractor, supplier, and customer" has been replaced by "supplier, organization, and customer."

The most obvious change in the new standard is that there are no longer twenty requirements. The requirements have been regrouped as described in the following sections.

Quality Management System

- General requirements
- Documentation requirements: general, quality manual, control of documents, control of records

Management Responsibility

- Management commitment
- Customer focus
- Quality policy
- Planning: quality objectives, quality management system planning
- Responsibility, authority, and communication: responsibility and authority, management representative, internal communication
- Management review: general, review input, review output

Resource Management

- General
- Provision of resources
- Human resources: general, competence, awareness and training, infrastructure, work environment

Product Realization

- Planning of product realization
- Customer-related processes: determination of requirements related to the product, review of requirements related to the product, customer communication
- Design and development: planning, inputs, outputs, review, verification, validation, control of changes
- Purchasing: purchasing process, purchasing information, verification of purchased product
- Production and service provision: control of production and service provision, validation of processes for production and service operations, identification and traceability, customer property, preservation of product
- Control of monitoring and measuring devices

Measurement, Analysis, and Improvement

- General
- Monitoring and measurement: customer satisfaction, internal audits, monitoring and measurement of processes, monitoring and measurement of product
- Control of nonconforming product
- Analysis of data
- Improvement: continual improvement, corrective action, preventive action

By comparing the ISO 9001:2000 version with the current previous version of ISO 9001, it is apparent that the major new requirements include customer focus, internal communication, measurement and monitoring, customer satisfaction, and continual improvement. In other words, the company management must be actively involved in making the quality system work and in improving its effectiveness. The measurements and monitoring of the effectiveness of internal audits and corrective actions, the monitoring of processes, and the monitoring of products will address the weakness in the current previous implementation of ISO requirements.

One of the main ISO 9001 requirements is to identify key business processes and their critical characteristics for effectiveness, monitoring, and improvement. These critical characteristics must be working at the Six Sigma level to achieve superior profitability results. The requirement for continual improvement is described in ISO 9001:2000, as below:

> The organization shall continually improve the effectiveness of the quality management system through the use of the quality policy, quality objectives, audit results, analysis of data, corrective and preventive actions and management review.

The general requirements for a quality management system include identification of business processes through process mapping, sequence, and interaction of processes; criteria and methods for process controls; and the monitoring, measuring, and analysis of these processes. This is similar to the DMAIC (define, measure, analyze, improve, and control) elements of the Six Sigma methodology. Table 15.1 identifies similarities between ISO requirements and the DMAIC methodology.

Management (CEO or president) must accept the fact that the quality system is not a quality department's activity anymore. As with Six Sigma, the quality system is instead a chief executive's tool to manage business for growth and profitability. The management must understand that the quality system addresses all aspects of business operations. However, if the operations are not managed as intended in the ISO 9001 standard, the chances of profitability are already quite slim.

Integrating Six Sigma methodology into the continual improvement section will reduce the cost of managing the Six Sigma program. Accordingly, the Six Sigma methodology must be institutionalized, audited for effectiveness, and reviewed for performance. Risks for not managing the Six Sigma through ISO 9001 system include loss of focus, higher cost of Six Sigma, and a less process-based Six Sigma approach. Projects based purely on Six Sigma methodology may lead to interfunction rivalry and suboptimal performance.

TL 9000 (Telecommunications Industry)

TL 9000 quality system requirements, which are similar to the ISO 9001:2000 requirements, have been specifically designed for suppliers of the

Table 15.1 Similarities of ISO Requirements and DMAIC Methodology

Six Sigma Methodology	ISO 9001 : 2000 Requirements	Similarities
Define	4.1 General requirements 4.2 Documentation requirements	Both require identification of key business processes, measurements of effectiveness, and clear definition of processes to manage and improve.
Measure	8.1 Measurement of customer satisfaction, processes, and products	Both ISO 9001 and Six Sigma methodology require measurements.
Analyze	8.4 Analysis of data	ISO 9001 and Six Sigma require collection and analysis of data. The Six Sigma methodology requires more dramatic improvement to produce better bottom-line results.
Improve	8.5.1 Continual improvement	ISO 9001 requires continual improvement to positively influence profitability, while Six Sigma requires profitability-related improvement.
Control	7.5.1 Control of production and service provision	ISO 9001 in general is about process control through documentation, implementation, analysis, and improvement. The Six Sigma methodology emphasizes use of statistical techniques.

telecommunications industry. The goals of TL 9000 are to:

- Foster quality system efficiency and effectiveness
- Establish a common set of requirements
- Simplify quality system requirements
- Define cost- and performance-based measurements
- Drive continuous improvement
- Enhance customer satisfaction

The TL 9000 requirements include customer involvement and long-term and short-term planning for key business performance measurements such as the following:

- Cycle time
- Customer service
- Training
- Cost
- Delivery
- Product reliability

The training requirement includes training for Quality Improvement Concepts and Advanced Quality Training. The Quality Improvement Concepts include training for fundamental concepts of quality improvement, problem solving, and customer satisfaction. The Advanced Quality Training includes training in statistical techniques, process capability, statistical sampling, data collection and analysis, problem identification, problem analysis, and improvement.

To meet the quality improvement requirements, an organization must implement an improvement program to improve customer satisfaction, product quality and reliability, and processes. TL 9000 specifically identifies field performance data to drive internal quality improvement. The Six Sigma methodology, again, is a great approach to achieve the continual improvement requirements of TL 9000.

AS 9100 (Aerospace Industry)

The basic intent of the AS 9100 quality system is to achieve steady and long-term improvement in products and processes by continually improving the performance of design, manufacturing, administration, and support processes. The AS 9100 system has been developed by the international aerospace industry to provide uniform standards for aerospace suppliers, and it offers the opportunity for aerospace companies to implement a more effective, value-added quality system. Its requirements are very similar to those of ISO 9001:2000, with some additional requirements to improve product reliability and configuration management. However, AS 9100 includes identification of key characteristics that are critical to quality according to the design or contract requirements. Besides continual improvement requirements, AS 9100 identifies statistical techniques such as DOE, failure mode, and effect analysis. AS 9100 encourages the use of tools such as lean thinking, process mapping, team leadership, project management, and process improvement. The Six Sigma methodology would be helpful in the aerospace industry while complying with the AS 9100 requirements.

The Six Sigma methodology would heighten the awareness of quality improvement that is needed to perform perfectly at 35,000 feet or higher. Throughout the supply chain in the aerospace industry, the Six Sigma methodology would bring benefits as a result of sensitivity to quality improvement and business performance.

QS-9000 and ISO 16949 (Automotive Industry)

Quality standards for the automotive industry are in alignment with the Six Sigma methodology in the sense that they both link to business performance. The QS 9000 standards have requirements for continual improvement and business planning. QS 9000 also requires development of suppliers in implementing quality practices. The automotive industry is known to be a very demanding industry that has required its suppliers to pursue continuous improvement aggressively. Some automakers have even started implementing Six Sigma methodology to achieve significant process improvement. The QS-9000 requirements, such as customer business plan, analysis and use of company level data, benchmarking, FMEA, team approach, and use of statistical techniques, all are elements of the Six Sigma methodology. With its current level of experience with statistical approaches, Six Sigma is a natural step in making the quality system more effective and achieving a higher rate of improvement.

Given competitive pressures to perform better, faster, and cheaper, an organization must plan to perform better than customer expectations of price reduction. A company sustained purely on the basis of price reduction will be reduced to nothing. Therefore, value-added and innovative thinking must be institutionalized to achieve higher profitability objectives. Without profitability, it would be difficult for any organization to create higher capabilities and increase its bottom line. Interestingly, auto industry executives realize the market demands and have already taken the initiative to implement Six Sigma or equivalent methodology to survive and strive.

The ISO/TS 16949 standard was developed by the International Automotive Task Force (IATF), Japan Automobile Manufacturers Association Inc., and ISO/TC 176. The ISO 16949 is an internationally recognized quality management system standard equivalent to the QS-9000 standards. It has been aligned with the ISO 9001:2000 requirements. The goal of the ISO/TS 16949 is to promote continual improvement, emphasizing defect prevention and reduction in variation and waste in the supply chain. Again, requirements for knowledge of statistical techniques, business planning, and benchmarking support implementation of the Six Sigma methodology to reduce cost, improve customer satisfaction, and improve profitability.

ISO 14001 (Environmental Management System)

The environmental management system ISO 14001 consists of requirements including environmental policy, planning, implementation and operation, checking and corrective actions, and management review. In comparison with the ISO 9001 system, the environmental management system produces faster results due to its focus on operations and waste reduction. The ISO 14001 standards were developed to enable an organization to formulate a policy and objectives regarding significant aspects of operations. These environmental aspects, if improved at the rate observed in the Six Sigma methodology, would lead to direct impact on waste reduction and, therefore, on profitability. According to ISO 14001, the organization shall establish environmental objectives and targets for each relevant function and level within the organization. These objectives and targets must be considered while planning for Six Sigma implementation. Once the targets have been established, an organization develops an action plan to achieve the desired targets. Experience shows that organizations have established targets that could be easily achieved, in order to maintain compliance to the standards.

These environmental aspects and targets are great opportunities to define projects for implementation of the Six Sigma methodology. By applying the focus and urgency of Six Sigma projects, the organization is bound to achieve a higher rate of improvement.

IPC 9191 (Electronics Industry)

IPC, a leading professional organization in the electronics industry, has developed standards for process controls and implementation of Six Sigma in the electronics industry: IPC-9191, General Guidelines for Implementation of Statistical Process Control. This standard specifies guidelines for use when an organization needs to achieve the capability to reduce variation in processes associated with design, or reduce variation in processes associated with design of production. The standard requires a company to have the management philosophy, commitment, and planning to implement process improvement and controls throughout the organization.

The objectives of process controls according to the IPC-9191 standard are as follows:

- To increase knowledge about the process
- To steer a process to behave in the desired way

- To reduce variation of final-product parameters, or in other ways improve performance of a process

The basic steps to implement process controls, such as strategic planning, training, identification of key process characteristics, statistical controls, and monitoring company-wide in an organization are similar to the steps that are required in implementation of the Six Sigma methodology. However, the IPC-9191 standard is focused more on manufacturing in the electronics industry. The Six Sigma methodology applies throughout operations and creates opportunity for dramatic results in manufacturing as well as nonmanufacturing areas.

MBNQA (MALCOLM BALDRIGE NATIONAL QUALITY AWARD)

The Baldrige award was created to accelerate the pace of innovation and to improve business performance. The award criteria provide a framework to assess performance based on key business indicators such as customer, product and service, financial results, human resources, and operations. The award criteria have been a valuable tool to align resources, improve communication, productivity, and effectiveness, and achieve strategic goals. Studies have shown that organizations that implemented systems based on the award guidelines outperformed their counterparts. Since its inception, the award has been instituted in healthcare and educational institutions.

The criteria are designed to help organizations use an integrated approach to organizational performance management that results in the delivery of ever-improving value to customers, contributing to marketplace success and the improvement of overall organizational effectiveness and capabilities. Core values of the criteria include the following:

- Customer-driven excellence
- Managing for innovation
- Management by fact
- Focus on results and creating value
- Systems perspective

Reducing defects and errors, meeting specifications, and improving customer service contribute to customer-driven excellence. An organization's ability to recover from mistakes is equally important in retaining customers. Customer-driven excellence is directed toward customer retention, market share gain, and growth. Customer-driven excellence is about creating value that the customer loves. Achieving the highest value requires a well-educated approach to organizational and personal learning. Organizational learning includes both continuous improvements of existing approaches and adaptation to change, leading to new goals and/or approaches. Table 15.2 identifies the differences between various performance levels.

The focus and intensity of the Six Sigma training, commitment to implementation, and selection and recognition of Black Belt candidates is such that Black Belts are expected to produce great results. Expectations that each project will lead to savings of about $250,000 per year, and that each Black Belt will lead four to six projects, require that the Black Belts be able to produce great results.

The rate of improvement and business growth requires innovation that makes breakthrough-level improvement in an organization's products and services. Both MBNQA and Six Sigma require dramatic results that are possible only through innovation in daily activities.

The MBNQA criteria (Table 15.3) include strategic planning, process management, information and analysis, and business results. Business decisions are made based on facts regarding customer satisfaction, product and service performance, and suppliers' and employees' performance. Successful management of the business performance requires a systems approach. Systems thinking means managing the whole organization, as well as its components, to achieve success.

The MBNQA criteria for performance excellence, as shown in Table 15.3, identify many areas that are addressed through the Six Sigma methodology. The Six Sigma methodology addresses customer focus, financial results, business processes, employee education and training, information analysis for problem definition, strategic planning, and leadership. It takes in passionate leadership, aligning Six Sigma with other business initiatives to achieve financial results through human resources, information analysis, and customer satisfaction. Leaders such as Bob Galvin of Motorola, Jack Welch (formerly) of GE, and Larry Bossidy of Allied Signal have shown their leadership in pursuing the Six Sigma initiative and superior financial results at different times.

MBNQA Criteria for Healthcare and Educational Organizations

The MBNQA criteria for performance excellence were created for healthcare and educational organizations. The following tables show the MBNQA categories and point values that similar processes apply to healthcare and

Table 15.2 Personal Performance Levels

Level	Classification	Business Impact
1	Questionable	When an organization delivers incompletely or to an unacceptable level. The output is unsatisfactory, and the customer tells anyone he knows how upset he is about the product or service he has received from this organization. The business cannot survive with this performance.
2	Commendable	An organization delivers the product or service that meets customer-specified requirements. The customer believes a fair trade of value and compensation has occurred, and forgets the supplier. The business may not survive if the competition is much stronger.
3	Excellence	When employees in an organization are learning and put effort into delivering a product or service that exceeds customer requirements. Customers like the product and service. When someone asks for a referral for the product or service, the customer promptly refers to the supplier. The company is a good company; however, the company is continually improving.
4	Great Performance	In this case employees in an organization perform with total dedication, are knowledgeable, and even forget about the time when providing the product or service. The care, skills, and effort, combined, lead to a superior product or service that customers love. The customer tells everyone he meets how happy he is about the product or service and recommends the supplier. The company is growing rapidly based on its best-in-class products and customer care. To produce Six Sigma level, or virtually perfect, work requires great performance from an organization's employees. That is achieved through the leaders' passion to achieve great results.

Table 15.3 MBNQA Criteria for Performance Excellence

Categories	Areas	Point Values for Categories
Leadership	Organizational Leadership Public Responsibility and Citizenship	120
Strategic Planning	Strategy Development Strategy Deployment	85
Customer and Market Focus	Customer and Market Knowledge Customer Relations and Satisfaction	85
Information and Analysis	Measurement and Analysis of Organizational Performance Information Management	90
Human Resources Focus	Work Systems Employee Education, Training, and Development Employee Well-Being and Satisfaction	85
Process Management	Product and Service Processes Business Processes Support Processes	85
Business Results	Customer-Focused Results Financial and Market Results Human Resources Results Organizational Effectiveness Results	450

educational organizations. Six Sigma methodology has been implemented in healthcare organizations; however, educational organizations have not been aggressive in pursuing Six Sigma methodology. The core values of healthcare and educational organizations are in Table 15.4.

As with any business, healthcare and educational organizations have processes for services, business management processes, and support processes. Actually, errors in such organizations could be much more expensive. Errors in healthcare organizations could be life-threatening, while errors in educational organizations could be lifelong. The tenacity of the Six Sigma methodology and passion for excellence and service should lead to improvement in price per patient for healthcare and higher performance in our schools.

Table 15.4 Organizational Core Values Based on the MBNQA Criteria

Healthcare	Education
Delivery of ever-improving value to patients and other customers	Student learning results
	Student- and stakeholder-focused results
Improvement of organizational effectiveness and capability	Budgetary, financial, and market results
	Faculty and staff results
Organizational and personal learning	Organizational effectiveness results

CONCLUSION

Six Sigma has been reviewed in comparison to the other quality systems that are being employed across industries. It is apparent that, although the basic intent of these standards is similar to the Six Sigma methodology, they do not explicitly entail the passion to achieve significant results and profitability improvements, as is the case with Six Sigma. An organization must align various quality initiatives to maximize business performance improvement through the Six Sigma methodology.

16

FINAL THOUGHTS

Six Sigma deployment requires training the organization and committing to organizational excellence in order to achieve world-class performance and bottom-line profitability. Implementation of any change effort is difficult at best. The comprehensive tools and techniques of Six Sigma assist in sustaining lasting change based on the strategic plan and desired business results of an organization. Some reasons for business change include: desire for increased profits and increased market share, sustaining a loyal customer base, increased employee satisfaction, and greater organizational effectiveness and efficiency. There is strong evidence that unsuccessful change is the result of a lack of clear goals, failure to understand the need for change, poor leadership commitment and communication, lack of incentives for change, and allowing those resisting change to win.

Six Sigma deployment offers an alternative while at the same time providing major improvements. Those who wish to participate or lead their organization in the application of Six Sigma are guilty of professional ignorance if they do not enhance and continue to develop their personal leadership capability. Champions and Black Belts have direct and immediate leadership challenges. The Champions are charged with interfacing with senior management and the Black Belts. Their ability to gather support, resources, and commitment from senior management will often mean the difference between success and failure in a Six Sigma effort. If the Champions do not lead the Black Belt selection, education, and development, odds for success decrease dramatically. One of the most critical leadership responsibilities for the Champion is to ensure that Black Belts who have spent two years working on improvement projects are moved back into responsible positions within the organization. If this essential leadership responsibility is not actively pursued, then the organization will soon see the Six Sigma Black Belt position as a career-ending move. The organization severely limits the effectiveness of any Six Sigma effort if their Black Belts are not reassigned back into the line organization where their skills can be leveraged.

A shortsighted approach is to profit from the Black Belts' projects that have returned value to the bottom line of the organization, and attempt to maintain these people in that position. Some will argue that the Black Belt training was expensive in both time and money, and therefore Black Belts should remain in these positions beyond the recommended time. We consider this thinking to be a short-term view that will eventually keep the organization's best and brightest from accepting Black Belt assignments in the future. One might consider the advantage managers gain from the Black Belt experience. Organizations that require Black Belt service find that their best and brightest people will seek those positions.

Black Belts have a duty to increase their leadership capability while working on projects. Team selection and building offer the Black Belts excellent opportunities to develop leadership skills. Any Black Belts who are not increasing their knowledge and skill base in people leadership, project management, and understanding of the language of management (money) during each project are kidding themselves and the organization that has invested in them. Black Belts are given some wonderful opportunities to add to their technical skill training, enhance their leadership skills, and apply those skills on projects in a very rapid fashion. Any organization that views Black Belts as technical experts is missing the largest benefit of Six Sigma.

Senior management that is not selecting its managers from the ranks of those who have been trained and practiced as Black Belts does not understand the true benefit leverage of a Six Sigma implementation. Within a few years after starting a Six Sigma application, no person should be assigned to a management position who has not completed Black Belt training and demonstrated mastery of the skills by successfully completing several projects. Six Sigma becomes the primary management development program within the organization. If this does not happen, then the senior management has had an unconscionable lapse in leadership application. In this case, Six Sigma deployment will not have the opportunity to deliver the kinds of improvements possible—not because the Six Sigma process cannot deliver the improvements, but because of the lack of leadership and the misapplication of this very powerful approach.

Management must ensure that the Six Sigma training and development afforded the Black Belts are appropriate for the high-performance expectations. While statistics understanding and application is very important, we consider the synergistic combination of behavior and leadership skills equally important in producing successful Black Belts.

Value Measurement

Measuring value is strictly a customer satisfaction issue. No matter what your product or service, customers determine with their dollars the value of your proposition. It is possible to change that understanding through a variety of efforts on your part. However, make no mistake: the customer's satisfaction is the ultimate determination of the value of your proposition. Six Sigma seeks to understand that value from the point of view of the customer, and to enhance the value in a way that is advantageous to both the customer and the organization.

Some organizations, because of unusual circumstances, find themselves in a high-sales environment with unusually high margins. Frequently, organizations convince themselves that the customer willingly agrees that there is adequate value involved in their product or service, and that the real reason for the high-sales environment with unusually high margins is some sort of marketplace discontinuity. That is not to say that there is not a lot of money to be made during these times.

"Pet Rocks" were a nice fad that made handsome profits for some people. As a fad, pet rocks died in a very short time. If your business seeks to capitalize on this sort of marketplace, you must realize that it is short-lived and that the value comes from the unique idea, not from some underlying value or need in society. Any new idea or product is in the position of trying to balance the true value and the "newness" component. A classic example of this is the early Ford Motor Company. Henry Ford insisted on driving the cost of cars down and making them inexpensive so that the market could become bigger and bigger, rather that selling fewer cars at much higher prices. Yet when consumer tastes evolved to wanting colors other than black, that was seen as a fad that would increase the cost and would soon disappear. What almost disappeared was the Ford Motor Company.

Few organizations can wait for the marketplace to inform them of the exact value of a product or service; even the Ford Motor Company barely survived a missed reading. Yet in many cases if too much time is spent on study and evaluation, the profitable opportunity may pass. Small organizations have the ability to react and change rapidly. A few large companies are able to keep their product pipeline filled with profitable innovation. Each year, 3M expects 40 percent of its sales revenue to come from products that are less than four years old. In some businesses it takes that long to get approval for model changes of the same product. One of the most important competitive advantages in the marketplace today is speed. Cycle time reduction, in addition to eliminating waste, has the real benefit of providing more opportunities to learn. Every cycle is a learning opportunity. Those who learn in every cycle soon reap a huge competitive advantage from completing the cycle one more time than their opponent in the same period, because they have gained knowledge.

Customer value will drive the redesign and evolution of your products and services. The customer's voice, combined with technical considerations, drives the product and service evolution. Those who are not staying ahead of this evolution will soon suffer. Videos are a good example. Betamax from Sony was first to market and, from what we are told, was a superior technical product to VHS. However, VHS format, coupled with inexpensive players and a marketing move to generate a variety of movies, drove the Beta Max from the field. One of the places that we believe Six Sigma has great untapped potential is in the marketing functions of most companies. When marketing, design, and manufacturing become team-focused on driving customer value, then breakthrough will generally be achieved.

One method of measuring value is to consider the defects in the entire process of delivery to the customer. Lower defect levels will be of higher value. With Six Sigma, the concept of rolled-through yield is very important. If defects occur in multiple places or stages before reaching the customer, they can all be added together to get the total number of defects per unit. Using a Poisson approximation:

$$\text{Yield} = e^{-\text{DPU}}$$

Note that if you average one defect per unit, the yield is 0.36788. Some have multiple defects, and about 36.7 percent will make it through without any defects. Anything that is less than the expectation of the customer is a defect. Doing some simple math, 63.3 percent of your customer interfaces will have a defect. Multiple surveys indicate that fewer than 4 percent of customers will ever complain. Four percent times 63.3 equals 2.532, or approximately 2.5 percent. If 2.5 percent of your customers (each transaction counted as a unit) complain, then you will have a defect rate averaging one per unit.

In the December 2000 issue of *Quality Progress*, Gregory H. Watson suggests that there are three logical categories for failures or defects. First, the product or service does not meet customer expectations; second, the price is not appropriate for the customer to see sufficient value; and third, the delivery is not within the required period for the customer. The assumption here is that all defects are the same in the eyes of the customer. This view of measuring defects from the perspective of the customer is essential if you are going to move the improvements, through Six Sigma, from incremental internal improvements to the dramatic customer-focused changes that are found with world-class performance. All of the non-product-related defects are considered in the yield. Here our yield is customer satisfaction. The implications of a 3 percent complaint rate by customers are terrible. Only by chance are you able to produce a defect-free product or service 36.7 percent of the time, and all of the remaining products or services have at least one defect, which means that some have multiple defects. Few businesses can survive with that level of poor performance. Measuring value is more than just considering the value of a product being delivered to the customer as intended. The value has to consider the total customer experience, including all of the support services such as logistics, accounts receivable, and so forth.

PRODUCT/SERVICE DESIGN AND REDESIGN

Most products or services that do not change and evolve will have limited success in the marketplace. Some businesses, such as the auto industry, have locked themselves into a cycle of a new model every year. Frequently these are minor or just cosmetic changes. With Six Sigma, the redesign of products and services should be based on the measured value from the customer's viewpoint, and on supplier capabilities. If done correctly, the redesigned product or service should do a better job of meeting and exceeding customer expectations than did the original product or service. When Six Sigma is applied, the economic return on investment should improve for the business unit involved. Better products with fewer defects, cost reduction projects, better understanding of the customer, improved workflow, employee commitment, supplier involvement, and reduced cycle times are but a few of the contributing areas. Good redesign should reduce complexity, have fewer defects, improve reliability, and offer the customer functionality that was not in the original product or service.

When these objectives are met, there is a direct cause and effect relationship between redesign and improved margins with larger market share. Once your customers have developed a low tolerance for defects, you gain a unique marketing advantage over your competitors. This is true provided that you continue to focus on the Customer's Critical Criteria and improvement processes. In these areas, it will be very difficult for a competitor to match the level of performance your customers have learned to expect from your business organization.

In the overall marketplace, failure rates of new and redesigned products remain high, despite extensive market research and other efforts. One reason for this failure is the poorly designed customer surveys that are frequently conducted.

Six Sigma focuses on the Customer's Critical Criteria (the important variables) that drive a successful redesign of a product or service. Understanding what the voice of the customer is really saying is of vital importance. The redesign of the product or service should do a better job of meeting the identified Customer's Critical Criteria, or address customer concerns and issues not met by the original product or service. Among those to be considered are expanded functionality, reduced cost, improved reliability, more attractive design, and the like.

Quality Function Deployment (QFD) and the resulting "House of Quality" chart is a tool that can be used to balance these often conflicting requirements. Multifunctional teams are used in the development, and the resulting matrix can serve as an important communications tool. The Customer's Critical Criteria are listed and ranked. Usually a comparison between the existing design and competitors' designs is included in the evaluation. Engineering requirements that are needed to meet the voice of the customer are compiled. The relationships between the customer requirements and the "engineering know-how" are shown in a relationship matrix. Objective measurements for each requirement are identified, and technical difficulty assigned. Positive and negative relationships between design requirements are determined, along with relative importance ratings. We should use this study to identify important issue(s) that can be sent forward in yet another iteration concerned with detailed design. Drawbacks to Quality Function Deployment include the amount of time and effort required, survey errors, and some of the subjective assessments that are made. This is generally why many people choose to use simpler approaches for a first attempt.

A simple cause-and-effect matrix can often accomplish a good deal of the desired results with much less expenditure of resources and effort. The outputs of a process are assigned an importance value, such as $1 =$ low, $5 =$ medium, and $9 =$ high. The customers of the process are asked to provide the ranking. Important process inputs are ranked on the same scale for their impact on each output variable. A team of process experts should achieve consensus on these rankings in a team meeting, independent of the customer rankings.

The two rankings are then placed in a matrix, with the process variables in the vertical column and the customer rankings in the horizontal row. The resulting product (multiplication of the row times the column) is placed in each cell.

Next cells in each row are added to create a Pareto Chart. This reveals which input variables have the biggest impact on the customer. This sort of matrix can be carried down to another level where the key process variable can be broken down into attributes, and the process broken down into the individual steps. This allows the process steps to be ranked according to the impact on the various attributes. We have found that a sort of stepwise analysis is easier for most team members to understand and accept than a Quality Function Deployment matrix. In addition, it can be carried out at several deeper levels.

DATA ANALYSIS

Any organization that wants to be successful must have a structured way of gathering data concerning all of the important areas and then turning that data into information. Raw data are not worth very much by themselves. It is only when the data are tortured to reveal information that they become valuable. This analysis of the data can vary in complexity and sophistication from very simple averages and histograms up to complex prediction models. Data collection is brought about for a number of reasons. In every area we have discussed, data are analyzed to produce information. When they are based on information, better decisions result at all levels in the organization. A few of the uses of data and information are discussed in the following sections.

Monitoring

Monitoring can include market conditions, performance of various parts of the organization, competitor actions/performance, and much more. When monitoring is executed correctly, the area being monitored has some level that prompts action. All of the statistical process control rules for special cause variation are examples of action criteria from monitoring. Dr. Edward Deming warned against tampering—that is, making special cause corrections for a system that is in statistical control. When a process is in statistical control, common cause variation is present, a system improvement is required, and an entirely different solution is needed.

A business strategy will have a number of essential performance areas, each with a number of metrics that the leadership wants to improve. It is always an interesting exercise to find out what is improvement, or which direction is good. Is there a targeted goal for each metric, and are we on target for achieving that goal? Multiperiod efforts that do not meet the target and come as a surprise to management are a clue that an adequate monitoring system is not in place. Measurements and performance indication should show whether progress is being made or not. Seldom do efforts push on for eleven months without change, and then suddenly produce all their benefits in the twelfth month. With proper reporting and tracking, Six Sigma projects should deliver in the areas important to the achievement of the strategic plan.

Benchmarking

One of the early steps in any effort is to understand the current performance. This can then be used as a benchmark against other organizations that have a similar process. A benchmarked point is also valuable in determining if activities and efforts associated with a Six Sigma project have had any impact. We caution against using a single-point value for making decisions. Frequently there are cyclic patterns, seasonal effects, and other influences that should be considered. At a minimum, we believe you should look at the average over some period and then compare that average with an average over a similar period after the project was implemented. Of course, the variation should also be considered. Six months of poor performance followed by one exceptional month can easily yield the same average as seven months at moderate performance levels, yet the information is much different.

A common flaw in benchmarking is to target a single performance metric. Most organizations do not have a clear understanding of the interrelationships involved, and can therefore improve one area to the detriment of another. An example of this is in the area of waste. It is relatively easy in many processes to change solid waste to air emission or liquid waste. Concentration on only one of these as a metric can often result in no net change in the total pounds released into the environment. The waste just changes form, with no net change in actual waste. When benchmarking for a Six Sigma project, collect data from multiple times and for several different metrics. At a minimum, know the historical average performance and variation.

Analysis

Other flaws are the reorganization efforts that eliminate staffing, especially in support functions, accompanied by large claims of savings. The remaining people in a different cost center are still doing the work, and often in a less efficient fashion. Activity-based accounting can burst the bubble of many of these pseudo-savings efforts. When the Purchasing Department is rationalized, and line people pick up the slack, it is not uncommon to find that productivity has decreased, and that the actual cost to do the purchasing activity has increased, rather than decreased as claimed. Without a system to accurately collect and analyze the data, mistakes of this type are common.

As a suggested activity to find out if this is a problem in your organization, go back five years and find out the total cost structure for a business unit. Next, document all of the improvement ideas and projects during that five-year period and the claimed savings (both single points in time and those that were to carry forward). Use a simple time value of money and see if you are actually realizing the benefits claimed from all of those projects. In one case a vice president of manufacturing claimed that if all the benefits from projects over a five-year period were real, he would need no raw materials, no energy, and no payroll, and would still be able to double production.

Six Sigma projects can make major improvements in service and staff functions—just be sure that the measurements capture what is really happening. It is easy to transfer costs from one function or location to another. Make certain your projects return the benefit to the organization as a whole.

Prediction

Another very valuable use of data is to make predictions. In fact, one of the main reasons for using statistics of any kind is to gain some capability to make predictions. Many organizations have very sophisticated prediction models for many facets of the business. The computer folks have added to our lexicon GIGO or "garbage in, garbage out." Any prediction model is only as good as the inputs. Sensitivity analysis of any model is highly recommended. In fact, some people have conducted designed experiments on purchased proprietary models to gain a better understanding of the factors and interactions.

As you make decisions, we encourage you to take to heart the mantra: In God we trust . . . all others bring data. Once the data are in hand however, they must be tortured with statistical analysis to force them to disclose information. Time, resources, and capability all influence the amount of torture you will inflict upon your data to get them to reveal their information. Only with information do you have a true competitive advantage.

CONCLUSION

Exceptional leadership leads companies to world-class performance. Managers must define and communicate the strategic visions of the organization. Leaders must sustain the organization's strategic direction by uncovering their Customer's Critical Criteria. Next, they must map out and understand all the processes of their products and services. Then they must evaluate the "current as is" versus the "should be" as compared to their Customer's Critical Criteria. The next important step is for managers to organize and prioritize sustainable improvement opportunities designed to meet or exceed the Customer's Critical Criteria. This becomes the battle plan for effective change that conduces to increased world-class performance and bottom-line profitability. One major key to obtaining successful implementation of Six Sigma methodologies is the alignment of the organization's visions, values, and systems. Forging these into strategic objectives, goals, and plans creates a vigilant focus on the activities and behaviors of the organization's efforts towards the achievement of world-class performance.

Certainly we have stressed that one must apply action to a clear and well-communicated organizational strategic plan. One such action requirement is to train the organization in the Six Sigma methodologies, tools, and techniques relevant to the current demands of the organization.

The purpose of this book is to provide organizations with a comprehensive look into the Six Sigma methodologies and implementations within various industries. Our aim is to provide practical information to facilitate successful implementation of the Six Sigma methodologies. Remember that Six Sigma is more than a process improvement or project improvement tool. When institutionalized, Six Sigma is part of a planned and monitored business strategy steered toward success. Our focus is on the dissemination of relevant improvement solutions for today's business environment. We hope you enjoy your journey and quest to learn more about how to deploy a Six Sigma initiative within your organization.

Use this book as a reference and guide as you devise your organization's Six Sigma deployment efforts. We wish you great success on your journey towards world-class performance and improved bottom-line profitability.

Appendix A

THE HISTORY OF SIX SIGMA

Since its innovation by the late Bill Smith of Motorola (Schaumburg, IL), Six Sigma concepts have been widely utilized in world industry. Bill doubtless never dreamed that one day his idea of measuring products and services would become an industry in itself. He would be proud of the recognition his breakthrough is receiving, but, at the same time, he would be amazed to know how the scope of Six Sigma has changed over time.

WHAT IS SIX SIGMA?

Since 1988, we have understood and trained others in Six Sigma methodologies. Six Sigma is a measure of the goodness of products and services, a philosophy, and a process. Higher sigma rankings mean better quality of a product or service, and lower sigma means poorer quality. Six Sigma initiative includes methodology steps as well as related methods and tools. Essential aspects are aggressive goal setting, graphical representation of performance against goals, effective quality management reviews, executive management expectations, standardized measurement systems, and inspiring leadership.

We wanted to learn about the current direction of the Six Sigma industry, so we ordered several Six Sigma books, attended a quality exposition to see various Six Sigma booths, encountered consultants, and collected various materials. We also learned that many of the publicized Six Sigma books are merely a collection of many quality tools that have been in existence for several generations. These quality tools are now considered tools of Six Sigma methodologies. Six Sigma is more than just the collection of tools. However, one must realize that the original Six Sigma was different from these trendy "Six Sigma" tools.

Some success stories, as well as the strong marketable leadership of a few organizations, have led Six Sigma to a pathway beyond imagination. In the late 1980s, we conducted our first process improvement experiment, using the variability reduction methodology and related simple quality improvement techniques, and achieved Six Sigma results.

Motorola grew dramatically between 1987 and 1992, the first five years of using the Six Sigma methodologies and tools. During this time, the company's sales doubled, profit margins improved, and Motorola's reputation soared. Those who worked in process improvement at Motorola did not throw dollars at Six Sigma to achieve this dramatic improvement. Instead, they coupled effective implementation of the methodology with a broader training program. We believe that simple practices, such as strong corporate leadership, good project management, and teamwork, helped achieve dramatic quality improvements at Motorola, improving the company's performance in general.

THE BIRTH OF SIX SIGMA

Let us travel back to 1981, the year Praveen joined Motorola's Semiconductor Products sector. Competition was intense, layoffs were feared, interest rates were 18 to 20 percent, and semiconductor chips were selling for less than they cost to manufacture. The market accused foreign companies of dumping chips, and to top it off, Motorola's financial reports were not looking very healthy in those years.

That era was the age of manufacturing 512 bits; 1K, 4K, and 16K dynamic random access memory (DRAM) chips were the latest technology. As with any state-of-the-art technology, the list of problems was continuous. However, Motorola was a leader in the communications industry, and the company's reputation was solid. Motorola was perceived as knowing how to manufacture effectively in the United States, while other companies were exporting their manufacturing offshore.

Although times were tough, Motorola would outperform competition during this particular economic downturn. However, Motorola's leadership realized a need for paradigm changes in order to survive the challenging times. Pressure to perform better, faster, and cheaper was strong, and Motorola's leadership set the company's quality improvement goal at ten times improvement over those five years (10 × 5).

With any new technology initiative, education for awareness and understanding is generally a first step. Motorola University was born as the vehicle to transform Motorola into a high-performance company. Managers attending this "university" would learn that change management created a work culture that adapted to a challenging and competing environment. A strong management review process was also established. Some managers would feel that going to their weekly management review meetings was like going to war because it was so thorough. During the first five years of this quality initiative, Motorola made significant progress; however, economic conditions were still challenging. Visionary leaders at Motorola, looking ahead through the maze of competition and manufacturing technologies, realized that this rate of quality improvement was not sufficient. As fast as manufacturers of calculators and watches saw their market eroding due to miniaturization and competition, Motorola's leaders could see personal telephones, televisions, two-way radios, and other products disappearing, unless quality improved dramatically and new products evolved.

Motorola's desire to be a leader in manufacturing in the core areas of communications, semiconductors, and industrial electronics led to a comprehensive benchmarking process and best practices. Phil Crosby's Zero Defects crusade was going on, with little success and considerable challenges in implementation. Its biggest obstacle was opposition to achieving perfection, because no one would commit to a perfect performance, likely due to a fear of failure.

Motorola's quality leaders were planning something more, and so from this concept we began to create a measurement method beyond three sigma, using normal distribution. Applying four sigma limits, the defective parts per million became 63 parts per million. Because the number of steps necessary to manufacture semiconductor chips was about 200, four sigma appeared to be a satisfactory level of quality. At four sigma, cumulative yield was about 99 percent, versus about 67 percent using standard three sigma limits.

Bill Smith, Communications Sector Quality Manager, had been contemplating a measurement method in a more realistic manner by allowing a shift of 1.5 sigma. The assumption is that processes behave normally and can fit Walter Shewhart's control chart theory. Accordingly, for a subgroup size of four, the control limits are set at 1.5 sigma for the individuals. Therefore, for a normally controlled process, if the mean shifts by 1.5 sigma or more, the process is out of control.

Limiting the maximum shift to 1.5 sigma of the process means makes it the same as the 3.0 limits of an individual point.

Bill had already established a correlation between field failures and internal failures, and concluded that most field failures were escaped internal failures, estimated to be in the ratio of ten to one (10:1). His dilemma was that, although manufacturing and design departments had improved their performance significantly, the results in product performance did not correlate. In addition, Motorola's benchmarking studies had shown that the customer expected almost a perfect product. With these two issues in mind, Bill conceived and integrated the Six Sigma measurements and design for manufacturability (DFM) methods. While developing these new methods, Bill had the total support of the Motorola leadership.

Once the Six Sigma concept was documented and an implementation strategy developed, the next step was the selection of projects. The initial projects had aggressive goals for improvement. The first four projects in a division, called Small Wins for Six Sigma, were in product development and manufacturing. Applying Six Sigma measurement methods established accountability that drove the improvement. The first departments to implement the Six Sigma methodologies were graphics and a new product development team.

On the cultural change side, all product managers were trained in Six Sigma and then trained their employees. As Six Sigma progressed, the quality expectation for each process and product was to obtain Six Sigma by achieving an annual quality improvement goal of 68 percent. The strong weekly, monthly, and quarterly quality reviews drove the performance to higher levels. Not every process or product achieved the 3.4 parts per million defect level, as it was a moving target due to changing customer requirements; however, the goal of implementing Six Sigma methodology at each process was significantly achieved.

Measurements were used to drive performance, employees' participation, leadership's interest, and higher awareness toward a common goal. A drive to achieve higher expectations regarding management and customers was an essential factor in institutionalizing the Six Sigma methodologies.

Motorola had announced to the world that it would achieve Six Sigma by 1992. By that year, the overall sigma level for the corporation was about 5.4 parts per million defects, a little less than Six Sigma. The amazing aspect was that a corporation of Motorola's size had successfully established a corporate-wide system to measure quality performance. Not many companies of any size have an integrated quality measurement system. Besides the measurement system, Motorola had institutionalized the philosophy of Six Sigma through the concepts of striving for perfection, creativity, dramatic improvement, and teamwork. Receiving the CEO award for achieving dramatic quality improvement from Bob Galvin was a great incentive for teams, over and above bonuses and other forms of recognition.

The first five years of Six Sigma at Motorola were very rewarding for the company. Sales grew dramatically. Introduction of better products with higher manufacturing capability and a focus on quality caused the company's reputation to soar. Outside Motorola, the employee badge could be as valuable as a credit card for identification. Such benefits of Six Sigma would not be realized by simply focusing on a few projects. Instead, Six Sigma was an organization-wide, leadership-driven, process-oriented, middle-manager-led, and employee-owned initiative.

Six Sigma has evolved over time. Today, Six Sigma requires a specific problem-solving project with significant potential return on investment or a bottom-line result. This approach has merit, and many consultants have converted their problem-solving approaches to Six Sigma methodologies, which requires serious senior management commitment to the methodology in dollars and time. Many executives wonder whether their company should commit to Six Sigma's problem-solving process. Six Sigma is not a silver bullet, and it is not a cure-all. Management should evaluate the potential benefits of initiating a Six Sigma deployment process.

IMPLEMENTING SIX SIGMA SUCCESSFULLY

Someone just learning the Six Sigma methodologies and transferring them to their business is likely to struggle. The following sections discuss various factors that made the Six Sigma methodology successful at Motorola, and can make it successful elsewhere.

Commitment

Motorola's success in implementing the Six Sigma processes required strong commitment from the executive management and senior management. At the management level, even Robert Galvin, Chairman of the Executive Committee, was talking about Six Sigma to Motorola customers, suppliers, and employees. Senior managers taught the class to their employees, to demonstrate to their people that they were committed to Six Sigma.

Common Language

Some wondered, "Why use Six Sigma, when using Defects Per Million Opportunities achieves the same results?" Translating DPMO to Six Sigma measurements provided a target, a clearer picture of the corporate goal, and a single direction to the company. Six Sigma became a corporate language and culture in which everyone talked about Six Sigma when it came to measuring process or product performance. Six Sigma crossed various divisions, products, functions, countries, and cultures. Soon, it became a common bond among Motorola's worldwide operations.

Aggressive Goals

One of the leadership aspects of the Six Sigma methodology is the desired rate of improvement required to reach the 3.4 parts per million goal. Typically, businesses set improvement goals at 10 to 20 percent per year in quality and other business measurements. The Six Sigma methodologies require improvement of 68 percent every year. This forces everyone to be creative in order to achieve the Six Sigma performances.

Innovation

Setting aggressive goals may be easy; however, realizing aggressive goals requires more than minor adjustment in processes. It really requires a totally new approach, new teamwork, and new thinking. This keeps people interested in what they are doing and allows them to achieve much better results. Setting incremental improvement goals makes people think that some tweaking will be sufficient to achieve those goals.

Process Thinking

There is some confusion about how to apply Six Sigma to different products, services, or functions. Six Sigma methodologies reinforce process thinking, as it requires analyzing the process. If the process is not changed, the gains will be short-lived.

Communication

Once it has been established as a common goal, Six Sigma is utilized universally within the company. It is publicized through newsletters, meetings, annual reports, quality policy cards, and of course via training. The organization's intent must be to train all employees so that they all have a common understanding of the Six Sigma processes.

Metrics

Six Sigma measurements must be utilized and acted upon throughout the organization. The process and product performance goals are set in terms of sigma. Accordingly, one modifies the procedures for process and process development to ensure effective implementation. Eventually, the Six Sigma processes are integrated into every aspect of business, including sales, purchasing, manufacturing, engineering, and management. An important aspect of measurements is to establish goals and measurements that will lead to a Six Sigma level of performance.

Improvement

Six Sigma methodologies promote the achievement of quality improvement. Expecting quality improvement, establishing quality improvement teams in various areas, reviewing quality improvement, and reporting the company-wide improvement are critical steps to achieve results. Executive management must be interested in learning about the process improvement at the process level as well as at the company level.

Rewarding Experience

Initially setting incentives to achieve "small wins" is a great way to get employees' attention and involvement. Stack the deck in favor of success. A journey to achieve Six Sigma is a celebration rather than a painful experience of punishment during the performance review. Quality awards, banquets, meetings, competition, suggestion programs, encouragement, and recognition are great ways to reward the Six Sigma experience, including success or partial success.

DESIGN FOR SIX SIGMA

After initial success, Six Sigma methodologies have become part of the strategy of many corporations. Before the creation of Six Sigma by Motorola, initiatives such as design for manufacturability (DFM), cycle time reduction (or lean manufacturing), and waste reduction existed. Six Sigma methodologies have evolved to incorporate DFM and lean manufacturing.

Quality assurance has evolved from end-of-line inspection (product inspection) to on-line inspection, to process control (control charts), and to off-line inspection (DFM), quality management systems (ISO 9000), and Six Sigma (dramatic improvement). The result of this evolution has been the invention of many tools, techniques, and systems. Commonly known are Pareto Charts, cause-effect diagrams, control charts, design of experiments, quality function deployment, Malcolm Baldrige National Quality Award Guidelines, value engineering, TRIZ (Russian methods for systematic innovation), ISO 9000, QS-9000, and Six Sigma. Names such as Walter Shewhart, Ed Deming, Joseph Juran, Armand Feigenbaum, and Phil Crosby are legends in the quality assurance industry. Motorola invented Six Sigma within a framework that included DFM and lean manufacturing.

Concurrent design of product and processes accomplishes DFM. Design for Six Sigma uses an integrated approach to design so that the product is manufacturable at the highest quality and lowest cost and satisfies all of the customer's requirements. DFM is important to implement because the majority of manufacturing defects are the result of design-related issues. The success in creating a manufacturable product depends upon clearly defined product goals reflecting the physical and functional requirements of the customer. Products designed for Six Sigma using design for manufacturability processes will allow the following:

- Virtually defect-free or robust product design
- Waste-free manufacturing
- 100 percent usable purchased parts
- Minimal maintenance and service
- Total customer satisfaction

In a design for Six Sigma environment, the product design team works with a cross-functional team consisting of members from marketing, sales, quality, manufacturing, and purchasing, and even with customers.

An excellent measure of a product designed for Six Sigma performances is Cp, defined as follows:

$$\text{Cp} = \frac{(\text{Upper specification limit} - \text{Lower specification limit})}{(6 \times \text{Standard Deviation})}$$

If the Cp is equal to or greater than 2.0, the product design can be considered a Six Sigma design because, when transferred to production, it will most likely yield 99.9996 percent for each customer's critical characteristics.

For Six Sigma designs, the product design team should focus on the following considerations:

- Fewest number of parts
- Parts of known capability
- Maximum design tolerances
- Maximum operating margins

Bill Smith, the inventor of Six Sigma methodologies at Motorola, has stated that electronics systems designed to the target conditions, given the above considerations and operating under normal conditions without overstress, will never fail.

Lean manufacturing is a way to specify value, arrange value-creating actions in the optimum sequence, conduct these activities without interruption, and improve continually. Lean thinking is a way to do more with fewer resources to provide customers with exactly what they want. Value is defined in terms of specific products with specific capabilities offered at a specific price to a customer. The value stream is a set of specific actions required for creating value. Identifying the value stream for each product or service is an essential step in lean thinking. Typically, we group the actions in the value stream in one of these three categories:

- Actions creating value
- Actions creating value but unavailable immediately
- Actions creating value and available immediately

Having identified the value stream, the wasteful steps are eliminated, and the remaining value-creating steps flow. Lean thinking is considered out-of-the-box thinking. Every step is questionable. One of the opportunities for change is reducing the batch size. Having a large batch size is a natural inclination for high-volume producers, to justify longer set-up times. Lean thinking challenges the batch size by focusing on reduction in set-up time. Batch size affects purchasing quantities, maintenance, set-up time, material flow, and quality improvement. One of the measurements of lean manufacturing is inventory level. Inventory is a good measure of manufacturing woes, because a natural company reaction is to build more or buy more just in case of a shortage. Suddenly, the cost of carrying the inventory starts eating up the profits.

Managing inventory is like managing a river, where the volume of water is dependent on the length, depth, and width of the river. Similarly, in a manufacturing operation, inventory level is dependent on the number of process steps (length), unique part counts in designs (width), and organizational policies (depth). When one starts reducing the inventory level, rocks—problems—start to appear. Continual improvement at a dramatic rate is a critical part of sustaining lean operations.

For parts manufactured with lean manufacturing and Six Sigma designs, manufacturing engineers should focus on the following considerations:

- Documentation
- Product design documentation
- Process manufacturing instructions
- Inspection and test procedures
- Repair and rework instructions
- Handling of nonconforming material
- Processes of known capability
- Simple and shortest process flow
- Process reproducibility
- Organizational policies

Products designed for Six Sigma and built with lean thinking are manufactured in a pull system and with virtual perfection. Ultimately, an organization's goal is to produce the highest quality product at the lowest cost and with minimal waste. In such an environment, endless improvement is realized using *kaizen* (continual), as well as *kaikaku* (dramatic), methods. Zero inventories with a batch size of one unit are the ideal.

Appendix B

TOOLS COMMONLY USED IN SIX SIGMA

One of the fundamental principles of Six Sigma is to solve problems, efficiently and effectively. The focus is on meeting and exceeding the Customer's Critical Criteria, which, as shown in Figure B.1, can be requirements and/or enhancers.

The purpose of the various tools and techniques is to allow you to better meet or exceed the Customer's Critical Criteria. Anything that fails to meet the Customer's Critical Criteria expectations is considered a defect. Defects can occur in the product or service, human interaction, advertising, delivery, accounts

Customer's Critical Criteria

- Requirements
  - Cost of doing business
  - Not even considered unless these are fully met
  - Different for each customer
- Enhancers
  - More than expected
  - May be personal preferences
  - After a period of time may become a requirement without being stated

Figure B.1 Customer's Critical Criteria.

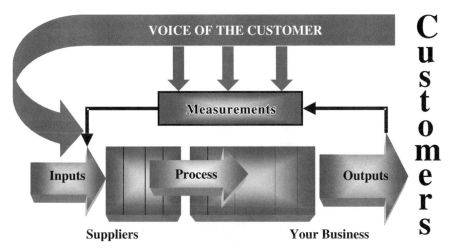

Figure B.2   Supplier and customer relationship.

receivable, sales, marketing, R&D, and so on. Everything in the organization has at least one opportunity for a defect, and most have many opportunities. Again, one of the fundamental principles of Six Sigma is to solve problems, efficiently and effectively. Figure B.2 shows the relationship of customers to our business and to our suppliers.

Defects can come from a variety of sources, essentially anywhere in the supplier-to-customer chain. There are significant benefits from eliminating the defects that are important to the customer and those internal to the organization. All too often these are not considered as part of the Cost of Poor Quality. Figure B.3 shows some of the areas for investigation for potential projects and for the application of the tools and techniques of Six Sigma.

Everything starts with the organizational strategy. As shown in Figure B.4, implementation of the strategic plan involves two separate implementation paths, both of which are essential for success:

- People development
- Improvement of processes and systems

One to the exclusion of the other will not be as effective as both in a synergistic approach. There are distinctly different tools and techniques for each. This relationship is shown in Figure B.4.

A strategic plan is the organizational *actions* in the *marketplace* that produce a *competitive advantage*. If all three elements are not present, then we

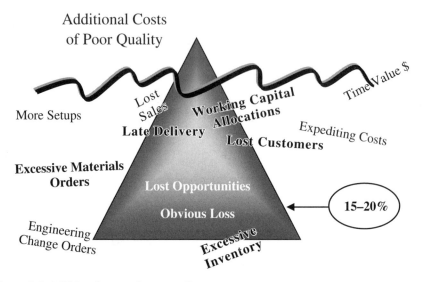

Figure B.3 Additional costs of poor quality.

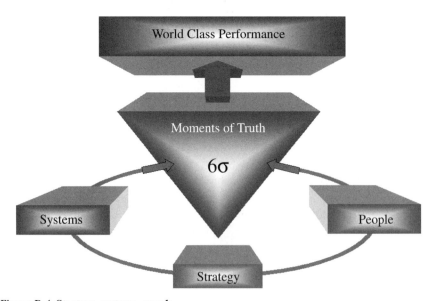

Figure B.4 Strategy, systems, people.

contend that a working strategic plan is not in place. There are many ambitious statements that may direct the development of a strategic plan, but until you can define the organizational actions you intend to implement, the marketplace where those actions will occur, and the competitive advantage that will result from taking those actions in the defined market, you do not have a strategic plan.

As with most things, tools and techniques implementations can be very simple or very complex. We recommend that, in all areas, you start with the simple approaches first and monitor the results. More complex tools and techniques can always be brought to bear on a specific situation as the need develops. Our maxim is: "Make the tools work for you, don't work for the tools." The objective is to get to a solution using a process that can be taught to a number of other people, so they can become better problem solvers themselves. There is no value added by using tools and techniques just to demonstrate that you can. The mark of the professional is to understand what tool or technique is needed for a given situation, and then to use it competently.

One of the benefits of Six Sigma is the structure provided for completing projects. While the tools may be used in a variety of places, much of the strength of Six Sigma is to have alignment with the strategic plan and then complete projects in a structured fashion. That project sequence is define, measure, analyze, improve, and control (DMAIC). There are opportunities to do people development at every phase.

Listed in Table B.1 are the various Six Sigma deployment phases and some of the tools and techniques that are commonly used during each phase. Many of these tools have applications in multiple areas, and the skill sets are transferable. This table is intended as a simple guide for your consideration; it is not intended to be a cookbook. Most of us will develop personal favorites, and elect to use them as a first choice for a given situation or circumstance. Do not allow your preferences to prejudice or blind you regarding the selection of other equally important Six Sigma tools or techniques.

Many have elaborated on the specifics of Six Sigma statistical tools and techniques. Our intent here is not to rehash the specifics, but rather to remind you that the tools and techniques are there to improve your successes as problem solvers. In and of themselves, none of the tools has any intrinsic value. Do not use these tools and techniques in isolation, but rather effectively couple them with your other training and experience, and with the project discipline of DMAIC. All of us have acquired formal education and vast work experience over the course of our careers. Combining this knowledge with the tools, techniques, and structure of Six Sigma will enhance our success as problem solvers, in the effort to deliver world-class performance. Additionally, Six Sigma Black Belt and Green Belt training provides people with a detailed level of competence and skill in using many of these tools and techniques as part of a project. As with any tool, it is possible for it to become "rusty" if not used. Most of us have heard the saying "Use it, or lose it."

Table B.1 Six Sigma Tool Applicability Matrix

Component	Applicable Tool or Technique	Expected Outcome
Strategic Planning	Values, beliefs, principles	General guidance
	Vision	Where the organization is headed
	Mission	What the organization does
	Vision elements	Key areas of focus for the current planning cycle
	Key performance indicators (KPI)	Important areas for improvement
	Objectives	General results for each KPI
	Goals	Objective measurements of desired KPI levels
	Plans	Details
	Affinity Diagram	Group common ideas
	Matrices	Multiple comparisons
	SWOT	Strengths, weakness, opportunity, threats
	Market analysis	Competitive advantages
People Development	Behavior profile	Understanding one's own and others' social styles, strengths, and weaknesses
	Versatility	Working with different styles, reducing interpersonal tensions, and improving productivity
	Team training	Phases of team growth: Forming, Storming, Norming, and Collaborating
	Skills training	Demonstrate competence
	Change equation	Provide understanding of one's own and others' personal values and beliefs
	Goal accomplishment	Goal setting and achievement process

Table B.1 Six Sigma Tool Applicability Matrix (continued)

Component	Applicable Tool or Technique	Expected Outcome
	Personal interest and values	Understanding individual and team interests
	Meeting management	Purpose, process, and payoff of planning, conducting, and documenting meetings
	Communication skills	Skills of listening, writing, speaking
	Situational leadership	Leader's and subordinates' styles of directing, coaching, supporting, and delegating
	Directing	Specific formative instruction for new employees or those new to the job function
	Coaching	Leadership skills of coaching those who are still developing job and personal competency
	Supporting	Leadership skills designed to build confidence in individuals who have demonstrated job competency
	Delegating	Leadership skills for decision making on an independent level
	Black Belt career path	Future leaders
	Recognition reward systems	Motivation, inspiration, esteem
Define	Project charter	Scope, boundaries, timelines and constraints, team
	High-level process map	Start/Stop, major process steps, customers, process name, process owner
	Customer's Critical Criteria	Things that matter to the customer
	Market research	Customer's Critical Criteria
	Feasibility and magnitude	Potential gain, estimate of time and effort

(continued)

Table B.1 Six Sigma Tool Applicability Matrix (continued)

Component	Applicable Tool or Technique	Expected Outcome
Define (*cont.*)	Is/Is Not	What is included and what is excluded from a project
Measure	Financial indicators	Cost, margin, ROI, comparison to standard
	DPU, DPMO	Defect rates per unit or opportunity
	Average and standard deviation	Mean and variation of data
	Cpk, sigma level, MSDT	Performance vs. specifications, includes mean and standard deviation
	Evaluation of measurement system	Accuracy and precision
	Yield	Throughput yield, rolled-through yield
	Control charts	Difference between Common Cause and Special Cause variation
	Customer's Critical Criteria	Actual performance comparison
Analyze	Identify key variables	Identify the important things that change
	Correlation	Things that change together
	Regression	Model parameters
	FMEA	Determine risk priority for potential failures or defects
	Fault tree	Why an incident or defect occurred
	Detailed Process Map	Document all steps in the process, reduce complexity
	Cause and effect (Fishbone diagram)	Potential causes classified as manpower, methods, machinery, and materials

Table B.1 Six Sigma Tool Applicability Matrix (continued)

Component	Applicable Tool or Technique	Expected Outcome
	Pareto	Rank by impact or contribution
	F ratio	Compare variances
	t-test	Means vs. a value
	ANOVA	Compare means, model fitting
	Cycle time	Length of time for all process steps
	Reliability engineering	Probability that the intended function is not performed for the time period specified
	Histograms	Unusual shapes or patterns in data
	Run charts	Data in time series for patterns, trends, etc.
	TRIZ	Learn from other disciplines.
	Relationship diagram	Root causes
	5 Why	Root causes
Improve	Gantt Chart	Progress vs. Plan
	Design of experiments	Optimization, key variables, interactions
	Evolutionary Operations	Iteration to better performance
	Set limits on Parameters	Limits to provide optimal performance
Control	Control charts	In statistical control
	Procedures	Standard work practices
	Audits	Confirm that improvements and results are still in place
	Storyboards	Share project experience with others
	Customer's Critical Criteria	Performance over time
	Presentation skills	Effective reporting and sharing of results

(continued)

Table B.1 Six Sigma Tool Applicability Matrix (continued)

Component	Applicable Tool or Technique	Expected Outcome
Control (*cont.*)	Training and testing	Demonstrated competence
	Cpk, sigma level, MSDT	Maintaining the gains
	Lessons learned	Project successes and learning opportunities

We recommend that, when assigning projects, Six Sigma Champions should review the tools and techniques used by a Black Belt on previous projects. One might consider this a sort of assessment of personal tools and techniques. Projects can be assigned that are likely to require those specific tools or techniques. This allows the Black Belt to continue building personal competence and confidence in Six Sigma deployment methodologies. Six Sigma Champions are challenged with grooming Black Belts to become more valuable problem-solving employees. This challenge can only be met effectively when the Six Sigma Black Belts are given opportunities to demonstrate their skills. More suggestions regarding tools and techniques can be found in the appropriate appendices in this book.

Appendix C

TRAINING CERTIFICATIONS

MANAGEMENT AND CHAMPIONS TRAINING

Management and Champions training involves a three-day, interactive decision-making work session for the leadership of the organization, covering topics as follows.

Why Six Sigma?

What is Six Sigma?

- Advantages
- Difference from and similarities to Total Quality Management

Scope of Commitment

- Management support
- Time
- Training
- Resources
- Maintaining an evergreen system

Value

- Project returns
- Cost of Poor Quality

Overall Approach

Strategy

- What are the specific elements of your strategy?
- What are the performance indicators for each element of the strategy?
- What is the history for each of these performance indicators?
- What is the goal for each performance indicator?
- If the performance indicator goals are reached, will you have achieved the strategic objectives?

People

- Levels of training commitment for Six Sigma
- Black Belts
- Green Belts
- Awareness
- Fit with other development effort

Systems and Process

- Identification of Key Work Processes
- Linkage between Key Work Processes
- Owners for Key Work Processes
- Stakeholders for Key Work Processes

Loyal Customers

- Customer prioritization
- Volume
- Profit
- New ideas
- Emerging areas

Estimates

Cost of Poor Quality

- As a percentage of sales
- Actual data

Estimate of Number of Black Belts

- Percentage of the workforce
- Number to achieve desired annual return
- At least one from each major department

Champions

- By business

Champions' Role

Selection of Black Belts

- Criteria
- Number

Selection of Projects

- Criteria
- Number
- Owner of selection process
- Project review

Career Planning for Black Belts

- Belong to the business
- 2–3 years is about max
- Messages sent to the organization

Training

- Systems, continuously improving

- "Evergreen" Black Belts (train Black Belts and replace those moved to other assignments)
- Awareness training for the business
- Green Belt

Systems

Training

- Understand and document major work processes

Coaching and Mentoring

- Black Belts
- Green Belts
- Previous Black Belts and Green Belts

Software

- Selection
- Availability
- Computer

Process Owners

- Role with Black Belt projects
- Maintaining improvements
- Cross-functional accountability

Accounting

- Tracking Six Sigma investment
- Tracking returns
- Individual project cost and returns

Six Sigma Tools

Problem-Solving Tools

- Overview of fundamental improvement tools

Statistics

- Measurement
- Control charts
- Design of experiments
- Reliability engineering

People Development

Behavior

- Social styles
- Goal achievement
- Teams

Project Management

- Basic tools
- Define, Measure, Analyze, Improve, Control (DMAIC)

Communications

- Written
- Oral
- Presentations
- Meeting management

Black Belts

- Training
- Certification
- Duration

Green Belts

- Training
- Certification

Final Project Reviews

- Process and tools used
- Results
- Finance verification

- Learning
- Close-out communications

Black Belt Modules

SIX SIGMA INTRODUCTION

With Six Sigma Introduction, participants begin to discover and identify the major components of world-class performance. The module defines Six Sigma, presents the key elements of strategy, explains the motivation for organizational change, defines the steps to building a personal Six Sigma Black Belt project, and identifies the cost of poor quality. It also includes several assignments.

Focus on Strategy

Discussions and assignments center around commodity, technology, quality, service, and customer strategy-driven focuses.

Project Guidelines and Selection

Discussions and assignments center around Six Sigma project selection, common problems, defining the scope of your personal project, team selection, developing teamwork, leadership, communications, identifying team members' roles, and how productive teams work.

Service (Nonmanufacturing)

This module, with included assignments, develops the skills for: identification of waste, cycle time reduction, service processes, service strategies, service projects, management responsibilities, performance measurements, auditing defects, and understanding products versus services.

Defects

Skills training and assignments include developing participant awareness of the customer's desires and Critical to Quality (CTQ) areas. Participants learn to use Customer's Critical Criteria (CCC) to focus the organization's efforts on improvement activities. Steps included for identifying defects per unit and defects per million opportunities.

Behavior

Participant discussions and assignments are focused on (1) the stages of individual and team growth as they apply to performance, (2) how paradigms affect projects, (3) gaining understanding of self and others, and (4) skills to reduce interpersonal relationship tensions to improve productivity.

Data

The module focuses on collecting and using data. Several data analysis tools and techniques are demonstrated, such as normal distributions, dependent variables, and independent variables. Participants will also gain an understanding of why Six Sigma is at a 3.4 parts per million defect level.

DISC Profile

This module is an introduction to the foundational skills of the basic behaviors of self and others. Included in this training are the stages of individual and team growth as they apply to performance, and how paradigms can affect projects. Participants gain an understanding of and healthy respect for the strengths and weaknesses different behavioral styles can bring to the team. The basic motivation behind this training is the acquisition of skill sets that contribute to improved productivity by reducing interpersonal relationship tensions. The module includes a personal individualized behavioral profile for each participant.

Distributions

The module focuses on identifying common distributions, other than the normal distributions. Participants learn how to use distribution characteristics.

Versatility

This module requires the individual completion of a personal profile of behavior, intended as a tool to gain understanding of self and others. This module focuses on building successful teams through the strengths of individuals and through the synergy of interpersonal relationships in collaboration.

Tests

The module's discussion focuses on correlation and regression, analysis of means (ANOM), and analysis of variance (ANOVA). The module includes assignments.

DISC Profile Feedback Module

This module is both an individual and a group debriefing of the results of the individual personal profiles of behavior and feedback. The focuses are to assist participants in gaining a greater understanding of self and others, to provide suggestions for improving interpersonal relationships and reducing relationship tensions, and to explain the steps for improving win-win productivity.

Personal and Team Empowerment

This module addresses both myths and truths of management styles, shared leadership, teamwork, empowered teams, creating high performance, and styles of decision making. The module includes a review and use of the quality process tools.

Leadership and Motivation

The module focuses on the skill sets for motivation of self and others. Included are recognition of self-limiting fears, using time management skills, awareness of performance ceilings, and tapping into individual motivational breakthrough. Participants discover how to use training, rewards, challenges, and consequences for developing an impassioned motivation leadership style.

Personal Project Work

Participants are required to work on assignments in specifically identified areas of their personal Six Sigma Black Belt project after each major block of training, prior to proceeding to the next major block of training.

Customer Service

This module focuses on assisting participants in discovering what the customer wants, and, as a direct result, how that knowledge plays into the organization's strategic plans. The module addresses how to decide who in the organization will be involved with the customer, what knowledge is typically required, and how to gain the necessary customer communications skills.

Creating a Customer Service Plan

This module focuses on basic customer feedback methodologies, as well as how the organization can measure customer satisfaction.

Fundamentals of Finance

This module assists the participant in gaining an understanding of finance as the language of business. This overview module explores the basic accounting equation, the double-entry system, income statements, balance sheets, and some common ratios.

Financial Controls

This financial overview module explores basic bookkeeping, cash flow, credit and collections, fixed assets, inventory, financial statements, profit and loss, balance sheets, financial analysis (profit, breakeven, and ratios), survival, cutting costs, financial reporting, assessments, and audits.

Project Measurement Evaluation

Six Sigma Black Belt project teams develop an evaluation of the measurement system for an ongoing initial project. This work includes the purpose of the measurement, the system involved, short-term stability, bias, and discrimination.

Budgets

This overview module helps participants gain an understanding of the "how" of defining a budget, and the methods for preparing budgets, allocations, and variances.

Repeatability and Reproducibility Studies

This overview module teaches participants how to conduct a study, develop necessary planning, and perform data collection and analysis.

Economic Evaluation

This module is an overview of economic evaluation, which includes methods for evaluating project economics. These methods include return on investment (ROI), cash flow, and payback.

Cycle Time Reduction

Participants of this module gain understanding of the advantages of cycle time reduction, the dynamics of the basic team structure, and visual tools regarding measurement, management, and proactive change.

Project Management Introduction

This module presents an overview of the basic project management system. Participants gain an understanding of the consequences of a structured project management system, as it relates to the accomplishment of the organization's desired business results.

Defining a Project

This module explores the scope, boundaries, expectations, justifications, and limitations of the project management process.

Planning a Project

Six Sigma Black Belts will use a variety of tools and techniques to plan their projects' critical path method (CPM), among others.

Project Budgets

This overview module helps participants gain an understanding of the "how" of defining a budget, and the methods for preparing budgets, allocations, and variances.

People for a Project

This overview module helps participants gain an understanding of their staffing/destaffing capabilities.

Project Execution

Participant discussions and assignments focus on completing the project and tracking the project's progress.

Project Reporting

Participant discussions and assignments focus on the interim reports, verbal reports, presentations, and written reports required of Six Sigma projects.

Project Close

Participant discussions and assignments focus on Six Sigma project completions, sharing learning experiences with the organization, and documentation processes.

Reading Control Charts

This module assists participants in understanding common and special cause variation, and investigation responsibility. Participants gain skills in reading different kinds of control chart signals.

X Bar R Charts

Participants learn how to construct, subsize, and interpret range and X Bar charts. The module includes assignments.

XmR Charts

Participants' discussions and assignments focus on XmR chart development, applications, and interpretations.

C Charts and u Charts

Participant discussions and assignments focus on how to construct, use, and interpret C charts and u charts.

p and np Charts

Participant discussions and assignments focus on how to construct, use, and interpret p and np charts.

Graphical Approach to Design of Experiments

This module provides participants with a graphical overview of the design of experiment process. The focus is on assisting participants in the understanding and appreciation of the scientific aspects of this analytical process.

Contrasts, Effects, Sum of Squares, Interactions

Participant discussions and assignments focus on calculations for the design of experiment process.

Models: Diagnosis and Interpretation

Participant discussions and assignments focus on understanding the model and interpreting the results.

Fractional Factorials

Participant discussions and assignments focus on how to get good information with less than all the possible trials.

Replication and Randomization

Participant discussions and assignments focus on the importance of repeating the trial randomization of the order.

Planning for Design of Experiments

This module focuses on Six Sigma Green Belt participation in team selection, variable identification, selection of the range, and iterative experiments when planning for design of experiments.

Root Cause Investigation

This module provides participants with processes and skills for investigating, organizing, correcting, and documenting the root causes and contributing factors of incident findings.

FMEA

This module presents to participants the methods and tools of analysis, discovery, potential failure modes of a system, effects of failure on a system, and corrective actions. FMEA is a tool of root cause analysis.

Fault Tree Analysis

This module focuses on a top-down approach to failure analysis, starting with an undesirable event, working downward through a logical step process, and determining ways a failure can happen. Fault tree analysis is a tool of root cause analysis.

Reviewing and Debriefing

This module focuses on the What, Why, and How of experiential learning and on ways to share the lessons learned for improvements and corrective actions.

Writing Procedures and Instructions

This module presents a few styles, methods, forms, and checklists for consideration when developing written procedures and/or instructions.

Audits

This module focuses on general audit guidelines, planning, sampling, nonconformance reports, internal and external audits, and corrective actions.

Integration with ISO 9000

This module examines the integration of Six Sigma and ISO 9000 processes.

Presentation Skills

This module focuses on making different kinds of formal presentations. Included are suggestions for presentation planning, delivery, visual aids usage, question-and-answer segments, and short-notice presentation changes. The module contains a checklist for planning individual as well as team presentations.

Inventive Problem Solving

This module and related assignments focus on the introduction to concepts of TRIZ (Russian theory of inventive problem solving).

Reliability Engineering Concepts

This module and related assignments focus on the basics of reliability engineering, Reliability growth, the use of the Weibull chart's failure rates and mean time between failures, and reliability block diagrams.

Procedures and Writing Specifications

This module presents a few styles, methods, forms, and checklists for consideration when developing written procedures and/or instructions.

Project Review

This module's focus is a review of the Six Sigma Black Belt's personal project. It includes the detailed project elements of define, measure, analyze, improve, and control.

GREEN BELT TRAINING

Green Belt training is designed for managers, Six Sigma project team members, and Green Belt project leaders. Green Belt competence is dependent on individual, organization, and project needs.

Six Sigma Introduction

The module identifies the major components of world-class performance, defines Six Sigma, identifies the key elements of strategy, and explains the motivation for organizational change. Also covered are the costs of poor quality and the steps for building a personal Six Sigma project.

Project Guidelines and Selection

This module covers Six Sigma project selection, common problems, defining the scope of your project, team selection, teamwork leadership, communications, team members' roles, and how productive teams work.

Service (Nonmanufacturing)

Participants learn how to identify waste and the skills of cycle time reduction. Also covered in this module is the identification of service processes, service strategies, and selecting service projects. The course helps participants to identify management responsibilities, proper selection of performance measurements, auditing defects, and issues of products versus services.

Defects

This module assists participants in understanding and identifying their customer's desires and CTQ areas of concern or interest. Participants discover how to use the CCC requirements and expectations to vigilantly focus the organization's improvement initiatives. Participants begin to explore the concepts of identifying defects per unit and defects per million opportunities and minimum loss.

Behavior

This module is an introduction to the foundational skills of understanding the basic behaviors of self and others. Included in this training are the stages of individual and team growth as they apply to performance, and how paradigms can affect projects. Participants gain an understanding of and healthy respect for the strengths and weaknesses different behavioral styles can bring to the team. The basic motivation behind this training is the acquisition of skill sets that contribute to improved productivity by reducing interpersonal relationship tensions. The module includes a personal individualized behavioral profile for each participant.

DISC Profile

This module requires the individual completion of a personal profile of behavior, intended as a tool to gain understanding of self and others. This module focuses on building successful teams through the strengths of individuals and through the synergy of interpersonal relationships in collaboration.

Personal Behavioral Profile Feedback Module

This module is both an individual and group debriefing of the results of the individual personal profiles of behavior and feedback. The focuses are to assist participants in gaining a greater understanding of self and others, to provide suggestions for improving interpersonal relationships and reducing relationship tensions, and to explain the steps for improving win-win productivity.

Skills of Versatility

This module is the last in the series regarding the foundational skills of behavior. The module deals with perceptions of self and others. Participants discover how these perceptions can affect the dynamics of both personal and team relationships and the productivity of said relationships. The module's focus is on improving one's ability to meet the needs and expectations of others by making them comfortable with one's behavioral style. The module addresses how to attain "win-win" interpersonal relationships, using skills of synergy and collaboration.

Data

The module focuses on collecting and using data. Several data analysis tools and techniques are demonstrated, such as normal distributions, dependent variables, and independent variables. Participants will also gain an understanding of why Six Sigma is at a 3.4 parts per million defect level.

Personal and Team Empowerment

This module addresses both myths and truths of management styles, shared leadership, teamwork, empowered teams, creating high performance, and styles of decision making. The module includes a review and use of the quality process tools.

Leadership and Motivation

This module focuses on the skill sets for motivation of self and others. Included are recognition of self-limiting fears, using time management skills, awareness of performance ceilings, and tapping into individual motivational breakthrough. Participants discover how to use training, rewards, challenges, and consequences for developing an impassioned motivation leadership style.

Customer Service

This module focuses on assisting participants in discovering what the customer wants, and, as a direct result, how that knowledge plays into the organization's strategic plans. The module addresses how to decide who in the organization will be involved with the customer, what knowledge is typically required, and how to gain the necessary customer communications skills.

Creating a Customer Service Plan

This module focuses on basic customer feedback methodologies, as well as how the organization measures customer satisfaction.

Fundamentals of Finance

This module assists the participant in gaining an understanding of finance as the language of business. This overview module explores the basic accounting equation, the double-entry system, income statements, balance sheets, and some common ratios.

Financial Controls

This financial overview module explores basic bookkeeping, cash flow, credit and collections, fixed assets, inventory, financial statements, profit and loss, balance sheets, financial analysis (profit, breakeven, and ratios), survival, cutting costs, financial reporting, assessments, and audits.

Project Measurement Evaluation

Six Sigma Green Belt project teams develop an evaluation of the measurement system for an ongoing initial project. This work includes the purpose of the measurement, the system involved, short-term stability, bias, and discrimination.

Budgets

This overview module helps participants gain an understanding of the "how" of defining a budget, and the methods for preparing budgets, allocations, and variances.

Economic Evaluation

This module is an overview of economic evaluation, which includes methods for evaluating project economics. These methods include ROI, cash flow, and payback.

Cycle Time Reduction

Participants of this module gain understanding for the advantages of cycle time reduction, the dynamics of the basic team structure, and visual tools regarding measurement, management, and proactive change.

Project Management Introduction

This module presents an overview of the basic project management system. Participants gain an understanding of the consequences of a structured project management system, as it relates to the accomplishment of the organization's desired business results.

Defining a Project

This module explores the scope, boundaries, expectations, justification and limitations of the project management process.

Reading Control Charts

This module assists participants in understanding common and special cause variation, and investigation responsibility. Participants gain skills in reading different kinds of control chart signals.

Graphical Approach to Design of Experiments

This module provides participants with a graphical overview of the design of experiments process. The focus is on assisting participants in the understanding and appreciation of the scientific aspects of this analytical process.

Planning for Design of Experiments

This module focuses on Six Sigma Green Belt participation in team selection, variable identification, selection of the range, and iterative experiments when planning for design of experiments.

Root Cause Investigation

This module provides participants with processes and skills for investigating, organizing, correcting, and documenting the root causes and contributing factors of incident findings.

FMEA

This module presents to participants the methods and tools of analysis, discovery, potential failure modes of a system, effects of failure on a system, and corrective actions. FMEA is a tool of root cause analysis.

Fault Tree Analysis

This module focuses on a top-down approach to failure analysis, starting with an undesirable event, working downward through a logical step process, and determining ways a failure can happen. Fault tree analysis is a tool of root cause analysis.

Writing Procedures and Instructions

This module presents a few styles, methods, forms, and checklists for consideration when developing written procedures and /or instructions.

Audits

This module focuses on general audit guidelines, planning, sampling, nonconformance reports, internal and external audits, and corrective actions.

Presentation Skills

This module focuses on making different kinds of formal presentations. Included are suggestions for presentation planning, delivery, visual aids usage, questions and answers segments, and short-notice presentation changes. The module contains a checklist for planning individual as well as team presentations.

ASQ SIX SIGMA BLACK BELT CERTIFICATION

Six Sigma certifications were developed at companies such as GE, Allied Signal, Motorola, and others over the past decade. The Six Sigma Academy has created their own Six Sigma Process and commercialized the Six Sigma breakthrough methodology. After its global success, the breakthrough methodology was adopted by many corporations. With increase in demand, many consulting corporations developed equivalent Black Belt training programs. That led to a question of core competency requirements for Black Belt certifications and tests to assess competency.

The American Society for Quality (ASQ), working with a few industry professionals, developed their own test. The test would assess understanding of concepts and applications of various techniques, to evaluate candidates' competency. The ASQ certification for Six Sigma Black Belt is called Certified Six Sigma Black Belt (CSSBB). The test provides an alternative to a three- to five-week-long training session, for quality professionals such as CQEs and those experienced in applying various statistical and improvement tools.

With the ASQ certifications, quality professionals have a choice to either go for a formal training offered by many resources, or take a test to qualify for certification. In addition to the test, the candidates must possess experience of at least two projects, or one project and equivalent experience, as stated in the requirements: "Six Sigma Black Belt requires two completed projects with signed affidavits or one completed project with a signed affidavit and three years work experience within the Six Sigma Body of Knowledge."

ASQ's Certified Six Sigma Black Belt offers an alternative to weeks of training. Some employees recognize and value the ASQ certifications as credentials that give them confidence in the knowledge and skills of those who have been certified. Many job ads for senior quality positions now mention the ASQ CSSBB as a required or desired credential.

Body of Knowledge (BOK) of CSSBB

The Body of Knowledge (BOK) for the SSBB is comprehensive. Six Sigma is built upon skills learned by a quality professional over ten years of experience. The skills include leadership, strategic planning, team building, graphical data collection and analysis tools, process improvement, design of experiment techniques, lean manufacturing, design for manufacturability, and innovative problem solving. The BOK also includes data-driven strategies for achieving target values and reducing variation.

Key Elements of Body of Knowledge

Module I. Enterprise-Wide Deployment

- Value of Six Sigma, business systems and processes, process inputs, outputs, and feedback
- Leadership, leadership roles in the deployment of Six Sigma, Six Sigma roles and responsibilities
- Organizational goals and objectives, key metrics/scorecards, project selection process
- Risk analysis; strengths, weaknesses, opportunities, threats (SWOT); scenario planning

Module II. Business Process Management

- Process vs. functional view, process elements, project measures
- Voice of the customer, critical customer requirements
- Business results, PPM, DPMO, DPU, RTY, COPQ, benchmarking

Module III. Project Management

- Project charter and plan, planning tools
- Team leadership, dynamics and performance, team performance evaluation, team tools
- Managing change, organizational roadblocks, negotiation and conflict resolution, communication, planning Tools

*Module IV. Six Sigma Improvement Methodology and
Tools—Define*

- Project scope, top-level process maps, metrics
- Problem statement, baseline and improvement goals

*Module V. Six Sigma Improvement Methodology and
Tools—Measure*

- Process analysis and documentation, tools, process inputs and outputs relationships
- Descriptive and inferential statistics, central limit theorem and sampling distribution of the mean
- Probability concepts, types of data, properties and applications of probability distributions
- Measurement systems, repeatability and reproducibility, metrology
- Process capability studies, process performance vs. specification, Cp, Cpk, Pp, Ppk, Cpm, sigma levels

*Module VI. Six Sigma Improvement Methodology and
Tools—Analyze*

- Multivariable studies, measuring and modeling relationships between variables, regression analysis
- Hypothesis testing, significance level, Type I and Type II errors
- Sample size determination, confidence intervals, ANOVA
- Tests for means, variances, and proportions; goodness-of-fit tests; contingency tables
- Nonparametric tests, Mood's median, Levene's test, Kruskal-Wallis, Mann-Whitney

*Module VII. Six Sigma Improvement Methodology and
Tools—Improve*

- Design of experiments terminology, planning and organizing experiments, design principles
- Design and analysis of one-factor experiments, full-factorial experiments, two-level fractional factorial experiments, Taguchi designs, significance of results, mixture experiments
- Response surface methodology, CCD, Box-Behnken, evolutionary operations (EVOP)

Module VIII. Six Sigma Improvement Methodology and
Tools—Control

- Statistical process control, rational subgrouping, control charts, PRE-control, short-run SPC, EWMA, measurement system re-analysis

Module IX. Lean Enterprise

- Lean concepts; theory of constraints; value, value chain, flow, pull system
- Continuous flow manufacturing, non-value-added activities, cycle time reduction
- Lean tools, total productive maintenance (TPM)

Module X. Design for Six Sigma (DFSS)

- Quality function deployment (QFD), robust design and process requirements, FMEA
- Tolerance design, tolerance and process capability
- Design for manufacturability, inventive problem solving (TRIZ)

Preparation for the Test

In order to prepare for the exam, one needs to assess a gap between the level of understanding required for each item and familiarity with the material. Having decided to take the examination, one should focus heavily on statistics application (SPC, DOE), lean manufacturing areas, and design for Six Sigma. One can even attend a review class to get a grasp of the BOK. Eventually, one needs to gather textbooks on Six Sigma and some detailed material that focuses very heavily on the statistical applications in Six Sigma. One must plan for about two months preparing for the test, reading through all this material (about 1,000 pages of material), which takes 150–200 hours. A good strategy for preparation is to become comfortable with the BOK, and bring an easy-to-use set of reference material. One must be familiar with the reference material and must become fluent in using it.

The breakdown of questions in the CSSBB examination can be seen in Table C.1.

The CSSBB exam encompasses an extensive body of knowledge that can be gained through experience over years. For professionals who have been

Table C.1 CSSBB Exam

Section #	Section Description	Number of Questions
1	Enterprise-Wide Deployment	9
2	Business Process Management	9
3	Project Management	15
4	Six Sigma Improvement Methodology and Tools—Define	9
5	Six Sigma Improvement Methodology and Tools—Measure	30
6	Six Sigma Improvement Methodology and Tools—Analyze	23
7	Six Sigma Improvement Methodology and Tools—Improve	22
8	Six Sigma Improvement Methodology and Tools—Control	15
9	Lean Enterprise	9
10	Design for Six Sigma	9
Total	**10 sections**	**150 Questions**

through training and projects, the CSSBB provides confidence that their Six Sigma Black Belt experience has been objectively recognized and certified by an independent body using a standard yardstick. The ASQ CSSBB is like a diploma that offers global acceptance for the body of knowledge and experience of a Black Belt.

Appendix D

BRIEF OVERVIEW OF SOME BLACK BELT PROJECTS

RAIL CAR CYCLE TIME

Define: Eliminate paying extra demurrage charges on rail cars.

Measure: Paying over four days' demurrage on some rail cars. Any demurrage charge over allowed is a defect.

Analyze: Rail car traffic, switch engine schedule, rail company operating rules, operating company procedures, spotting procedures.

Improve: Changed sequences of handling empty and full cars. Modified loading times by less than two hours. Result is essentially no demurrage, over the allowed, for the entire site.

Control: Rail company changed procedures, and operating company changed scheduling practices.

CHEMICAL PLANT BOTTLENECK

Define: Distillation tower has internal damage, limiting production rates. Next outage is scheduled in one year. If outage is taken now to repair damage, we will still have to take outage in one year because of parts delivery for other essential projects.

Measure: At anything over 85 percent of capacity, the distillation tower will not perform. With six months of effort, operations engineers and process engineering could find no solution other than to take an early outage. Anything less than 100 percent capacity is considered a defect.

Analyze: Identified key operating variables, established allowable ranges for each, and conducted a designed experiment.

Improve: A single set of conditions allowed operations at 102 percent of capacity without problems. At that level another part of the plant became the bottleneck. Increased capacity until scheduled outage was worth $6 million.

Control: All shift operators were trained for new conditions, and the operation's procedures were modified.

RETAIL DISPLAY

Define: Marketing has designed a "fancy" display unit that they think will outperform the "standard" display unit, and they want to put one in every store. "Fancy" display is ten times the cost of a "standard" display, and all stores already have "standard" units. Should the new displays be purchased?

Measure: Have data for each store on sales of this product for every day.

Analyze: The stores identified at least three other factors besides display type that could affect sales. Range for each factor was identified. Design of experiments was conducted.

Improve: "Fancy" display had no significant impact on sales. The "fancy" displays were not ordered for any more stores, with considerable cost savings.

Control: Future changes will be tested and evaluated using statistical techniques.

WATER TREATING

Define: In fifteen years, the water-treating unit had never been able to handle the nameplate capacity. Treatment chemical costs were higher than other types of treatment units.

Measure: Confirmed flow rate through the system, compared to nameplate.

Analyze: Measured system evaluation and found many measurements that were off by over 100 percent. Hourly operations identified key variables in the operation of the unit and the acceptable range of each. Conducted three different designed experiments.

Improve: Corrected the measurement problems. Found set of operating variables that produced 107 percent of nameplate capacity at higher quality with lower chemical use. Chemical use reduced by $180K per year.

Control: Hourly operations trained, procedures modified, process to check measurement instituted. Introduced model for changes in inlet water conditions.

POWER DISTRIBUTION RELIABILITY

Define: Large chemical site had significant losses due to power outages.

Measure: Dollar value determined for each failure and the total. Each failure was assigned to a major component.

Analyze: Mapped the entire system by major component and identified failure rates for each major component. Found areas with projects scheduled that were very unlikely to fail, and thus the improvement projects would add nothing to overall reliability. Other components with a high likelihood of causing an outage were being ignored.

Improve: Developed plan for each component, depending upon failure mode and frequency for that component. Made a tenfold reduction in the dollar losses due to power failures on site.

Control: Track each major component and modify action plan based on failure mode, if needed. System shared with other locations.

REDUNDANT ANALYSIS

Define: Analysis is being conducted at two and three locations for the same product, with different results from each location. Capital requests from multiple areas for the same analysis for the same material.

Measure: For each analysis, collected the corresponding results from each location. Totaled the capital request for analysis where they were already being done, or duplicate requests for the same analysis.

Analyze: In some cases the methods were the same and the brand of instrument the same; some had the same type of instrument but a different brand and different procedures; in others, different types of instruments were being used. Found overcalibration of most instruments. Sources of variation for each type of analysis were investigated using nested design of experiments.

Improve: Real-time telemetry of data eliminated some redundancy. For other analysis, correlation curves had to be developed to show the equivalent values for different methods, and agreement was reached to use one analysis and share the results. Totally eliminated the significant capital request for analysis.

Control: Modified capital authorization request procedure. Control charts for each analysis to determine when to calibrate.

NEW CAPACITY JUSTIFIED

Define: Contract to deliver product at a minimum rate on a daily basis. Severe penalties if rate missed by even a small amount. Customer "good will" also an issue.

Measure: Capacity of units in the system more than the minimum rates. Collected failure rate data for each unit and time to repair.

Analyze: Failure rate data combined with the time to repair data indicated that there were significant periods of time when the minimum contract rates could not be met, and penalties would be paid.

Improve: Capital approved for an additional unit. Within the first year, the new unit was required at least four separate times for several weeks, each time to meet the contract minimums. Any one of the four times returned enough cash to pay for all of the capital expended.

Control: System to track and monitor failure data and repair time data.

PEOPLE SELECTION

Define: Why is there such a difference in the sales performance of different people?

Measure: Top people have ten times the volume of the bottom 25 percent. Failure to meet sales quotas is a defect.

Analyze: Education, training, time in job, product line, sales area, profiles.

Improve: Able to identify by profile 72 percent of the top salespeople. Use this tool to select new people for this function.

Control: Use profiles for new hires, and continue to monitor performance levels.

PARTS FAILING AFTER FINAL MACHININGS

Define: Inspection is rejecting a high number of parts after final machines.

Measure: Product yield and number of defects in total were determined, to establish defect yield and sigma value.

Analyze: Machine operators, engineers, and vendor identified variables that could impact the production of defects. Range of acceptable levels determined for each variable. Five different designed experiments were conducted.

Improve: Operating instructions changed to the conditions with the lowest defect production consistent with capacity limits. Final product yield increased 13 percent.

Control: Control charts installed for each machine. Decision tree corrective action plan provided for known defects and known corrective actions.

OUT OF SPEC PRODUCT

Define: High percentage of product is out of spec and being automatically removed. No recycle or salvage value.

Measure: Quantified the amount of out of spec product for each product grade.

Analyze: Operations and engineers identified the variables that impact the production of out of spec material. Several of these are preventive actions performed by operations. Ranges for the levels and frequencies for the variables were determined. Designed experiments were run, and acceptable levels and frequencies determined.

Improve: Levels for the variables and frequencies for operator preventive actions established. Out of spec material dropped by 50 percent.

Control: Operating procedures were modified, schedules for operator corrective actions instituted, and control charts for the amount of out of spec material are being kept.

ENGINEERING CHANGES

Define: Large number of changes from client after approving engineering design. Schedule slipping.

Measure: Number of changes, time involved in changes, compliance to critical path schedule.

Analyze: No clear authority on client team to establish scope, any of client team could make changes, verbal communication of changes, conflicting changes by client team members. Language issues between client and engineers.

Improve: Regular engineering/client meetings where topics included scope for each section and desired objective, known limitations were defined, unclear requirements were questioned and options discussed. Written plan signed by client representative and engineering lead. Change requests in writing and signed by client representative. Changes decreased by factor of 4.7 and schedule was met.

Control: Change requests all in writing. Shared approach with other disciplines on project.

WEB DESIGN

Define: Design a Web site that ranks in the top ten on all major search engines and directories.

Measure: Enter "six sigma" and check ranking in search engines.

Analyze: URL name, title of pages, and other factors are major ranking criteria. Reciprocal links and other routine activities aid in search engine ranking.

Improve: Purchase URL with six sigma included, optimize each page, develop reciprocal links, and perform other regular activities required to maintain traffic and ranking.

Control: Monitor ranking on search engines weekly. You can check on the success of this project by entering "six sigma" in the search field of your favorite search engine. Success is a link to http://www.adamssixsigma.com in the top ten listings. The titles and descriptions may vary. The URL link is the performance measure.

Appendix E

GLOSSARY

5 "Why"s	Keep asking why at each successive level of detail.
Abscissa	The horizontal axis or line of a graph.
Accuracy	Closeness of agreement between an observed value and an accepted standard or reference value.
Activity	A process, step, function, or task that occurs over time. Combinations of activities form business processes.
Activity analysis	Analysis and measurement in terms of value, time, cost, and throughput of the steps that make up a process.
Activity-based costing (ABC)	Accounting technique that allows an organization to better determine the actual costs associated with each product and service produced by that organization without regard to the organizational structure.
Activity model (AS-IS)	A model that portrays how any process is currently arranged. It forms the baseline for future improvement.
Activity model (TO-BE)	A model developed as part of a business process redesign action or program. The TO-BE model shows how things are visualized after the improvements.
Activity, non-value-added	Anything that the customer would not pay for, or that is not required by law.

Activity, value-added	Something that the customer would pay for, or that is required by law.
Adequacy	Used in QS-9000. Indicates that the intent of the standard has been met, given the scope of the supplier's operation.
Affinity diagram	A tool used to organize ideas, usually generated through brainstorming, into groups of related families.
Algorithm	The rules for the solution of a problem, often a set of mathematical relationships or computer code.
Alpha risk	The probability of accepting the alternate hypothesis when, in reality, the null hypothesis is true.
Analysis of variance (ANOVA)	A basic statistical technique for analyzing experimental data. It subdivides the total variation of a data set into meaningful parts, each associated with specific sources of variation.
Apportionment	Assignment of reliability objectives from system to subsystem, such that the whole system will have the needed reliability.
Arrow diagram	Another term for a PERT or CPM chart.
Assignable cause	Variation that is not random.
Attribute	A characteristic that can take on only one value, e.g., A or B, good or bad.
Audit	An inspection to ensure that a process is conforming to its requirements.
Average chart (X-bar chart)	A control chart in which the average of the subgroup is represented by an X-bar. Average charts are paired with range charts or sample standard deviation charts.
Balanced Scorecard	Developed by Robert Kaplan and David Norton, a balanced business scorecard helps businesses evaluate how well they meet their strategic objectives. It typically has four to six components, each with a series of submeasures.
Baldrige Award	Malcolm Baldrige National Quality Award: An annual award given to an American company that excels in quality management and quality achievement.

Benchmark	The standard to which measurements or comparisons are made.
Best practice	The best way currently known of performing a task or process.
Beta risk	The probability of accepting the null hypothesis when, in reality, the alternate hypothesis is true.
Bias	A systematic error that contributes to the difference between a population mean of measurements or test results and an accepted reference value.
Black Belt	A Six Sigma project manager.
Block diagram	A simple flow chart showing basic relationships.
Boundary	A condition that limits a process or area of investigation. Start and stop points are boundaries.
Brainstorming	A method to get ideas from knowledgeable people. No criticism or discussion of ideas is allowed until all the ideas are recorded.
C charts	Charts that display the number of defects per sample.
Cause and effect diagram	Also called the fishbone chart or Ishikawa chart. The main line is the problem or issue, and the branches are commonly used methods, manpower, machines, and material.
Champion	A member of senior management who is responsible for the guidance and direction of Black Belts.
Characteristic	A definable or measurable feature of a process, product, or variable.
Central tendency	Mean, median, and mode.
Chart	A graphical display of information.
Charter	A document that specifies the purpose of a team, its power, its reporting relationships, and its specific responsibilities.
Checklist	A list of important steps that must take place, usually in the order in which they must be done. A list of things to do.

Check sheet	A form used to collect data. Usually, it is used to record how often some activity occurs.
Common cause variation	The normal variation within the process. Not an unusual event or condition, but not necessarily acceptable process performance.
Confidence levels	The probability that a random variable x lies within a defined interval.
Confidence limits	The two values that define the confidence interval.
Conformance	Meeting requirements or specifications.
Confounding	Two or more variables that vary together so that it is impossible to separate their unique effects.
Consensus	A decision that is acceptable to and supported by all.
Consumer's risk	The probability of accepting a lot that should have been rejected.
Continuous data	Numerical information where subdivision is conceptually meaningful and can assume any number within an interval, e.g., 14.7 psig.
Continuous improvement	Ongoing improvement.
Control chart	A chart that indicates upper and lower statistical control limits, and an average line.
Control plans	Control plans are written descriptions of what will be done to ensure that gains are maintained.
Corrective action	Action(s) designed to eliminate root causes.
Cost of poor quality	Internal and external failure cost plus appraisal and prevention costs.
Count chart (c chart)	An attributes data control chart.
Cp	Commonly used process capability index, defined as

$$[\text{USL(upper spec limit)} - \text{LSL(lower spec limit)}]/[6 \times \sigma]$$

where sigma is the estimated process standard deviation.

Cpk	Commonly used process capability index defined as the lesser of $$(USL - mean)/3\sigma$$ or $$(mean - LSL)/3\sigma$$ where sigma is the estimated process standard deviation.
CPM	Critical path method.
Customer	Any recipient of a product or service; also anyone who is affected by the product or service. A customer can be external or internal to the organization. Paying customers are external.
Cycle time	The time that elapses from the beginning to the end of a process. The time it takes to complete all the steps in a process, in the prescribed order, once.
Decision matrix	A tool used for issue evaluation. The possibilities are listed down the left-hand side of the matrix, and impact criteria are listed across the top. Each intersection is then rated. Can be categorical ranking or on a continuous scale.
Defect	Anything that does not meet the Customer's Critical Criteria.
Deming cycle	Alternate name for the Plan-Do-Check-Act cycle, a four-stage approach to problem solving. It is also sometimes called the Shewhart cycle.
Dependent variable	The response. For example, Y is the dependent, or "response," variable where $Y = f(X)$.
DFA	Design for assembly.
DFM	Design for manufacturing.
Discounted cash flow (DCF)	A method of performing an economic analysis that considers the time value of money.
Discrete random variable	A random variable that can assume values only from a definite number of discrete values.
DOE (design of experiments)	Intentionally varying certain factors by determined limits. Multiple changes are made at once.

DPU	Defects per unit.
DPMO	Defects per million opportunities.
Economic analysis	Costs and benefits of a proposal or project.
Empowerment	Giving employees decision-making and problem-solving authority within defined boundaries.
Expectations	Customer's Critical Criteria.
Factors	Independent variables.
Failure mode and effects analysis (FMEA)	A technique that identifies and ranks the potential failure modes of a design or manufacturing process, to prioritize potential improvement actions.
Fixed cost	A cost that does not vary with the amount or degree of production.
Flow chart	A graphical representation of a process.
FMA	Failure mode analysis; a tool of Root Cause Analysis.
Force field analysis	A method of representing forces, in which the driving forces are listed on the left with a right arrow, and the restraining forces are listed on the right with a left arrow.
Gage R&R	Gage repeatability and reproducibility.
Gantt chart	A horizontal bar chart that shows planned and completed work in relation to time. Each task in a list has a bar corresponding to it. The length of the bar is used to indicate the planned or actual time.
Green Belt	A person who has a demonstrated level of competence with Six Sigma philosophy, tools, and techniques. Green Belts are usually found in two roles: first in the management ranks, where they learn Six Sigma, and second on teams, where they acquire competence to participate on Six Sigma projects.
Histogram	Vertical display of a population distribution in terms of frequencies, where the width of the base is the interval, and the height is the number of occurrences.
Independent variable	A controlled variable.
Inputs	Products or services provided to a process.

Six Sigma	Structured application of tools and techniques on a project basis, to achieve sustained strategic results.
Six Sigma, application of	Define, measure, analyze, improve, and control steps.
Six Sigma failure rate	A failure rate of 3.4 parts per million, or 99.99966% good.
SOW	Statement of work.
SPC	Statistical process control.
Stratification	A process of grouping data according to a common characteristic.
Subgroup	A logical grouping.
Supplier	Anyone who provides inputs (materials, information, service, etc.) to a process. A supplier can be external or internal to the organization.
Theory	A plausible or general principle offered to explain phenomena.
TRIZ	A Russian theory of inventive problem solving.
Type I error	Rejecting something that is acceptable. Also known as an alpha error.
Type II error	Accepting something that should have been rejected. Also known as a beta error.
u chart	A control chart showing the count of defects per unit in a series of random samples.
Value-added activity	Something that the customer would pay for, or that is required by law.
Variance	A measure of deviation from the mean in a sample or population.
Variation	The differences among individuals.
Vision	A declaration often incorporated into an organizational statement to clarify what the organization hopes to be doing at some point in the future. The vision should act as a guide in choosing courses of action for the organization.
Voice of the customer	Customer feedback both positive and negative, including likes, dislikes, problems, and suggestions.

INDEX